Lecture Notes in Mathematics 2248

More information about this subseries at http://www.springer.com/series/3114

Fondazione C.I.M.E., Firenze

C.I.M.E. stands for *Centro Internazionale Matematico Estivo*, that is, International Mathematical Summer Centre. Conceived in the early fifties, it was born in 1954 in Florence, Italy, and welcomed by the world mathematical community: it continues successfully, year for year, to this day.

Many mathematicians from all over the world have been involved in a way or another in C.I.M.E.'s activities over the years. The main purpose and mode of functioning of the Centre may be summarised as follows: every year, during the summer, sessions on different themes from pure and applied mathematics are offered by application to mathematicians from all countries. A Session is generally based on three or four main courses given by specialists of international renown, plus a certain number of seminars, and is held in an attractive rural location in Italy.

The aim of a C.I.M.E. session is to bring to the attention of younger researchers the origins, development, and perspectives of some very active branch of mathematical research. The topics of the courses are generally of international resonance. The full immersion atmosphere of the courses and the daily exchange among participants are thus an initiation to international collaboration in mathematical research.

C.I.M.E. Director (2002 – 2014)
Pietro Zecca
Dipartimento di Energetica "S. Stecco"
Università di Firenze
Via S. Marta, 3
50139 Florence
Italy
e-mail: zecca@unifi.it

C.I.M.E. Director (2015 –)
Elvira Mascolo
Dipartimento di Matematica "U. Dini"
Università di Firenze
viale G.B. Morgagni 67/A
50134 Florence
Italy
e-mail: mascolo@math.unifi.it

C.I.M.E. Secretary
Paolo Salani
Dipartimento di Matematica "U. Dini"
Università di Firenze
viale G.B. Morgagni 67/A
50134 Florence
Italy
e-mail: salani@math.unifi.it

CIME activity is carried out with the collaboration and financial support of INdAM (Istituto Nazionale di Alta Matematica)

For more information see CIME's homepage: **http://www.cime.unifi.it**

Alexander Braverman • Michael Finkelberg •
Andrei Negut • Alexei Oblomkov

Geometric Representation Theory and Gauge Theory

Cetraro, Italy 2018

Ugo Bruzzo • Antonella Grassi • Francesco Sala
Editors

Springer

FONDAZIONE CIME ROBERTO CONTI
CENTRO INTERNAZIONALE MATEMATICO ESTIVO
INTERNATIONAL MATHEMATICAL SUMMER CENTER

Authors
Alexander Braverman
Department of Mathematics
University of Toronto
Toronto, ON, Canada

Michael Finkelberg
Faculty of Mathematics, Laboratory of Algebraic Geometry and its Applications
National Research University Higher School of Economics
Moscow, Russia

Andrei Negut
Department of Mathematics
Massachusetts Institute of Technology
Cambridge, MA, USA

Alexei Oblomkov
Department of Mathematics and Statistics
University of Massachusetts
Amherst, MA, USA

Editors
Ugo Bruzzo
Department of Mathematics
Scuola Internazionale Superiore di Studi Avanzati
Trieste, Italy

Antonella Grassi
Department of Mathematics
Università di Bologna
Bologna, Italy

Francesco Sala
Kavli Institute for the Physics and Mathematics of the Universe (WPI)
The University of Tokyo Institutes for Advanced Study
University of Tokyo
Kashiwa-shi, Chiba, Japan

ISSN 0075-8434 ISSN 1617-9692 (electronic)
Lecture Notes in Mathematics
C.I.M.E. Foundation Subseries
ISBN 978-3-030-26855-8 ISBN 978-3-030-26856-5 (eBook)
https://doi.org/10.1007/978-3-030-26856-5

Mathematics Subject Classification (2010): Primary: 14D20, 14D21, 14D22; Secondary: 14F05, 14J81, 81Q99, 81T13, 81T30, 81T45, 81T60

This Springer imprint is published by the registered company Springer Nature Switzerland AG.
The registered company address is: Gewerbestrasse 11, 6330 Cham, Switzerland

Preface

In the last 30 years a new pattern of interaction between mathematics and physics emerged, in which the latter catalyzed the creation of new mathematical theories. Most notable examples of this kind of interaction can be found in the *theory of moduli spaces*. In algebraic geometry the theory of moduli spaces goes back at least to Riemann, but they were first rigorously constructed by Mumford only in the 1960s. The theory has experienced an extraordinary development in recent decades, finding an increasing number of connections with other fields of mathematics and physics. In particular, moduli spaces of different objects (sheaves, instantons, curves, stable maps, etc.) have been used to construct invariants (such as Donaldson, Seiberg-Witten, Gromov-Witten, Donaldson-Thomas invariants) that solve long-standing, difficult enumerative problems. These invariants are related to the partition functions and expectation values of quantum field and string theories. In recent years, developments in both fields have led to an unprecedented cross-fertilization between geometry and physics.

These striking interactions between geometry and physics were the theme of the CIME School *Geometric Representation Theory and Gauge Theory*. The School took place at the Grand Hotel San Michele, Cetraro, Italy, in June, Monday 25 to Friday 29, 2018. The present volume is a collection of notes of the lectures delivered at the school. It consists of three articles from Alexander Braverman and Michael Finkelberg, Andrei Negut, and Alexei Oblomkov, respectively. Here we briefly summarize the contributions to this volume.

Braverman and Filkelberg's notes review the constructions and results presented in several joint papers with Hiraku Nakajima, where a mathematical definition of the Coulomb branch of 3D $N = 4$ quantum gauge theories (of cotangent type) is given. A section of the notes is dedicated to explaining the motivations coming from physics ("What do we (as mathematicians) might want from 3D $N = 4$ SUSY quantum field theory?"): Higgs and Coulomb branch and mirror symmetry, gauge theories. The lecture notes also include examples and explicit computations (in particular in the toric case). They also present a framework for studying some related further mathematical structures (e.g., categories of line operators in the corresponding topologically twisted theories).

The purpose of Negut's notes is to study moduli spaces of sheaves on a surface as well as Hecke correspondences between them. Hecke correspondences are varieties parameterizing collections of sheaves on a surface which only differ at points; they lead to interesting operators on the cohomology, Chow groups, and K-theory of the moduli spaces of sheaves. Their applications range from mathematical physics to classical problems in the algebraic geometry of hyperkähler manifolds. Oblomkov starts with an introduction to matrix factorizations and in particular equivariant matrix factorizations. Then he explains the homomorphism from the braid group to the category of matrix factorizations. He explains how one can construct interesting geometric realizations of the braid group, providing a categorification of the Markov trace. He also works out a computation of the above-mentioned trace for some small braids. The object of his lectures is related to the Kapustin-Saulina-Rozansky topological quantum field theory for the cotangent bundles to the Lie algebras as targets.

Trieste, Italy — Ugo Bruzzo
Bologna, Italy — Antonella Grassi
Kashiwa-shi, Chiba, Japan — Francesco Sala

Acknowledgments

We thank Alexander Braverman, Michael Filkelberg, Andrei Negut, and Alexei Oblomkov for their enthusiastic participation in the School, for delivering their lectures and writing up the notes. We also thank Roman Bezrukavnikov, Tudor Dimofte, Valerio Toledano Laredo, Oleksandr Tsymbaliuk, and Yaping Yang for delivering interesting talks, which were complementary to the courses offered. Thanks are due to the CIME Foundation, Compositio Mathematica, INdAM, and SISSA for the support. Special thanks are due to Elvira Mascolo and Paolo Salani, Director and Scientific Secretary of the CIME Foundation, for their help in the preparation of the CIME School and of this volume. And a final thank you to all participants and the staff of Grand Hotel San Michele for providing an excellent and enjoyable atmosphere.

CIME activity is carried out with the collaboration and financial support of "INdAM (Istituto Nazionale di Alta Matematica)". This school is partially supported by Foundation Compositio Mathematica.

Contents

Chapter 1
Coulomb Branches of 3-Dimensional Gauge Theories and Related Structures

Alexander Braverman and Michael Finkelberg

Abstract These are (somewhat informal) lecture notes for the CIME summer school "Geometric Representation Theory and Gauge Theory" in June 2018. In these notes we review the constructions and results of Braverman et al. (Adv Theor Math Phys 22(5):1017–1147, 2018; Adv Theor Math Phys 23(1):75–166, 2019; Adv Theor Math Phys 23(2):253–344, 2019) where a mathematical definition of Coulomb branches of 3d $N = 4$ quantum gauge theories (of cotangent type) is given, and also present a framework for studying some further mathematical structures (e.g. categories of line operators in the corresponding topologically twisted theories) related to these theories.

A. Braverman (✉)
Department of Mathematics, University of Toronto and Perimeter Institute of Theoretical Physics, Waterloo, ON, Canada

Skolkovo Institute of Science and Technology, Moskva, Russia
e-mail: braval@math.toronto.edu

M. Finkelberg
National Research University Higher School of Economics, Russian Federation, Department of Mathematics, Moscow, Russia

Skolkovo Institute of Science and Technology, Institute for Information Transmission Problems, Moskva, Russia

U. Bruzzo et al. (eds.), *Geometric Representation Theory and Gauge Theory*, Lecture Notes in Mathematics 2248,
https://doi.org/10.1007/978-3-030-26856-5_1

1.1 Introduction and First Motivation: Symplectic Duality and a Little Bit of Physics

1.1.1 Symplectic Singularities

Let X be an algebraic variety over $\mathbb{C}$. We say that X is singular symplectic (or X has symplectic singularities) if

(1) X is a normal Poisson variety;
(2) There exists a smooth dense open subset U of X on which the Poisson structure comes from a symplectic structure. We shall denote by ω the corresponding symplectic form.
(3) There exists a resolution of singularities $\pi : \widetilde{X} \to X$ such that $\pi^*\omega$ has no poles on $\widetilde{X}$.

This definition is due to A. Beauville, who showed that if condition (3) above holds for some $\widetilde{X}$ then it holds for any resolution of X.

1.1.2 Conical Symplectic Singularities

We say that X is a conical symplectic singularity if in addition to (1)–(3) above the following conditions are satisfied:

(4) X is affine;
(5) There exists a $\mathbb{C}^\times$-action on X which contracts it to a point $x_0 \in X$ and such that the form ω has positive weight.

We shall consider examples a little later.

1.1.3 Symplectic Resolutions

By a symplectic resolutions we mean a morphism $\pi : \widetilde{X} \to X$ such that

(a) X satisfies (1)–(5) above;
(b) $\widetilde{X}$ is smooth and π is proper and birational and the action of $\mathbb{C}^\times$ on X extends to an action on $\widetilde{X}$.
(c) $\pi^*\omega$ extends to a symplectic form on $\widetilde{X}$.

Example Let $\mathfrak{g}$ be a semi-simple Lie algebra over $\mathbb{C}$ and let $\mathcal{N}_\mathfrak{g} \subset \mathfrak{g}^*$ be its nilpotent cone. Let $\mathcal{B}$ denote the flag variety of $\mathfrak{g}$. Then the Springer map $\pi : T^*\mathcal{B} \to \mathcal{N}_\mathfrak{g}$ is proper and birational, so if we let $X = \mathcal{N}_\mathfrak{g}, \widetilde{X} = T^*\mathcal{B}$ we get a symplectic resolution.

1.1.4 The Spaces $\mathfrak{t}_X$ and $\mathfrak{s}_X$

To any conical symplectic singularity X one can associate two canonical vector spaces which we shall denote by $\mathfrak{t}_X$ and $\mathfrak{s}_X$. The space $\mathfrak{s}_X$ is just the Cartan subalgebra of the group of Hamiltonian automorphisms of X commuting with the contracting $\mathbb{C}^\times$-action (which is an finite-dimensional algebraic group over $\mathbb{C}$). The space $\mathfrak{t}_X$ is trickier to define. First, assume that X has a symplectic resolution $\widetilde{X}$. Then $\mathfrak{t}_X = H^2(\widetilde{X}, \mathbb{C})$ (it follows from the results of Namikawa that $\mathfrak{t}_X$ is independent of the choice of $\widetilde{X}$). Moreover, $\mathfrak{t}_X$ also has another interpretation: there is a deformation $\mathcal{X}$ of X as a singular symplectic variety over the base $\mathfrak{t}_X$. The map $\mathcal{X} \to \mathfrak{t}_X$ is smooth away from a finite union of hyperplanes in $\mathfrak{t}_X$.

If X doesn't have a symplectic resolution, Namikawa still defines the space $\mathfrak{t}_X$ and the above deformation; the only difference is that in this case $\mathcal{X}$ is no longer smooth over the generic point of $\mathfrak{t}_X$ (informally, one can say that $\mathcal{X}$ is the deformation which makes X "as smooth as possible" while staying in the class of symplectic varieties). More formally, if $\widetilde{X}^{\mathrm{reg}}$ stands for the smooth locus of a partial resolution ("as smooth as possible"), then $\mathfrak{t}_X = H^2(\widetilde{X}^{\mathrm{reg}}, \mathbb{C})$, see [33, Corollary 2.7].

1.1.5 Some Examples

Let X be $\mathcal{N}_{\mathfrak{g}}$ as in the example above. Then $\mathfrak{s}_X$ is the Cartan subalgebra of $\mathfrak{g}$ and $\mathfrak{t}_X$ is its dual space. One may think that $\mathfrak{t}_X$ and $\mathfrak{s}_X$ always have the same dimension. However, it is not true already in the case $X = \mathbb{C}^{2n}$. In this case $\mathfrak{s}_X$ has dimension n and $\mathfrak{t}_X = 0$.

Let now X be a Kleinian surface singularity of type A, D or E. In other words X is isomorphic to $\mathbb{C}^2/\Gamma$ where Γ is a finite subgroup of $SL(2, \mathbb{C})$. Thus X has a unique singular point and it is known that X has a symplectic resolution $\widetilde{X}$ with exceptional divisor formed by a tree of $\mathbb{P}^1$'s whose intersection matrix is the above Cartan matrix of type A, D or E. Thus the dimension of $\mathfrak{t}_X$ is the rank of this Cartan matrix. On the other hand, it is easy to show that $\mathfrak{s}_X$ is 1-dimensional if Γ is of type A and is equal to 0 otherwise.

1.1.6 The Idea of Symplectic Duality

The idea of symplectic duality is this[1]: often conical symplectic singularities come in "dual" pairs (X, X^*) (the assignment $X \to X^*$ is by no means a functor; we

[1]The main ideas are due to T. Braden, A. Licata, N. Proudfoot and B. Webster.

just have a lot of interesting examples of dual pairs). What does it mean that X and X^* are dual? There is no formal definition; however, there are a lot of interesting properties that a dual pair must satisfy. The most straightforward one is this: we should have

$$\mathfrak{t}_X = \mathfrak{s}_{X^*} \quad \mathfrak{s}_X = \mathfrak{t}_{X^*}.$$

Other properties of dual pairs are more difficult to describe. For example, if both X and X^* have symplectic resolutions $\widetilde{X}$ and $\widetilde{X}^*$ then one should have

$$\dim H^*(\widetilde{X}, \mathbb{C}) = \dim H^*(\widetilde{X}^*, \mathbb{C}).$$

(However, these spaces are not supposed to be canonically isomorphic). We refer the reader to [9, 10] for more details.

One of the purposes of these notes will be to provide a construction of a large class of symplectically dual pairs. Before we discuss what this class is, let us talk about some examples.

1.1.7 *Examples of Symplectically Dual Spaces*

1.1.7.1 Nilpotent Cones

Let $X = \mathcal{N}_{\mathfrak{g}}$ and let $X^* = \mathcal{N}_{\mathfrak{g}^\vee}$ where $\mathfrak{g}^\vee$ is the Langlands dual Lie algebra. This is supposed to be a symplectically dual pair.

1.1.7.2 Slodowy Slices in Type *A*

For partitions $\lambda \geq \mu$ of n, let $\mathcal{S}^\lambda_\mu$ be the intersection of the nilpotent orbit closure $\overline{\mathbb{O}}_\lambda \subset \mathfrak{gl}(n)$ with the Slodowy slice to the orbit $\mathbb{O}_\mu$. Then $\mathcal{S}^\lambda_\mu$ is dual to $\mathcal{S}^{\mu^t}_{\lambda^t}$.

1.1.7.3 Toric Hyperkähler Manifolds

Consider an exact sequence

$$0 \to \mathbb{Z}^{d-n} \xrightarrow{\alpha} \mathbb{Z}^d \xrightarrow{\beta} \mathbb{Z}^n \to 0$$

of the free based $\mathbb{Z}$-modules. It gives rise to a toric hyperkähler manifold X [7]. Then X^* is the toric hyperkähler manifold associated to the dual exact sequence (Gale duality).

1.1.7.4 Uhlenbeck Spaces

$\mathrm{Sym}^a(\mathbb{A}^2/\Gamma)^\vee \simeq \mathcal{U}^a_G(\mathbb{A}^2)/\mathbb{G}^2_a$ for a finite subgroup $\Gamma \subset SL(2)$ corresponding by McKay to an almost simple simply connected simply laced Lie group G. Here $\mathcal{U}^a_G(\mathbb{A}^2)$ is the Uhlenbeck partial compactification of the moduli space of G-bundles of second Chern class a on $\mathbb{P}^2$ equipped with a trivialization at infinity $\mathbb{P}^1_\infty \subset \mathbb{P}^2$, see [12]. Note that $\mathbb{G}^2_a$ acts on $\mathbb{A}^2$ by translations, and hence it acts on $\mathcal{U}^a_G(\mathbb{A}^2)$ by the transport of structure.

1.1.8 3d N = 4 Quantum Field Theories and Symplectic Duality

One source of dual pairs (X, X^*) comes from quantum field theory. We discuss this in more detail in Sect. 1.4; here we are just going to mention briefly the relevant notions.

Physicists have a notion of 3-dimensional $N = 4$ super-symmetric quantum field theory. Any such theory $\mathcal{T}$ is supposed to have a well-defined moduli space of vacua $\mathcal{M}(\mathcal{T})$. This space is somewhat too complicated for our present discussion. Instead we are going to discuss some "easy" parts of this space. Namely, the above moduli space of vacua should have two special pieces called the Higgs and the Coulomb branch of the moduli space of vacua; we shall denote these by $\mathcal{M}_H(\mathcal{T})$ and $\mathcal{M}_C(\mathcal{T})$. They are supposed to be Poisson (generically symplectic) complex algebraic varieties (in fact, the don't even have to be algebraic but for simplicity we shall only consider examples when they are). They should also be hyper-kähler in some sense, but (to the best of our knowledge) this notion is not well-defined for singular varieties, we are going to ignore the hyper-kähler structure in these notes. But at least they are expected to be singular symplectic.

There is no mathematical classification of 3d $N = 4$ theories. However, here is a class of examples. Let G be a complex reductive algebraic group and let $\mathbf{M}$ be a symplectic representation of G; moreover we shall assume that the action of G is Hamiltonian, i.e. that we have a well-defined moment map $\mu : \mathbf{M} \to \mathfrak{g}^*$ (this map can be fixed uniquely by requiring that $\mu(0) = 0$). Then to the pair $(G, \mathbf{M})$ one is supposed to associate a theory $\mathcal{T}(G, \mathbf{M})$. This theory is called *gauge theory with gauge group G and matter* $\mathbf{M}$. Its Higgs branch is expected to be equal to $\mathbf{M}/\!\!/\!\!/G$—the Hamiltonian reduction of $\mathbf{M}$ with respect to G. In particular, all Nakajima quiver varieties arise in this way (the corresponding theories are called *quiver gauge theories*).

What about the Coulomb branch of gauge theories? These are more tricky to define. Physicists have some expectations about those but no rigorous definition in general. For example, $\mathcal{M}_C(G, \mathbf{M})$ is supposed to be birationally isomorphic to $(T^*T^\vee)/W$. Here $T^\vee$ is the torus to dual the Cartan torus of G and W is the Weyl group. The above birational isomorphism should also preserve the Poisson structure.[2] In addition $\mathcal{M}_C(G, \mathbf{M})$ has a canonical $\mathbb{C}^\times$-action with respect to which the symplectic for has weight 2. Unfortunately, it is not always conical but very often it is. Roughly speaking, to guarantee that $\mathcal{M}_C(G, \mathbf{M})$ is conical one needs that the representation $\mathbf{M}$ be "big enough" (for reasons not to be discussed here physicists call the corresponding gauge theories "good or ugly"). In the conical case physicists (cf. [17]) produce a formula for the graded character of the algebra of functions on $\mathcal{M}_C(G, \mathbf{M})$. This formula is called "the monopole formula" (in a special case relevant for the purposes of these notes it is recalled in Sect. 1.5.3).

The idea is that at least in the conical case the pair $(\mathcal{M}_H(\mathcal{T}), \mathcal{M}_C(\mathcal{T}))$ should produce an example of a dual symplectic pair. One of the purposes of these notes (but not the only purpose) is to review the contents of the papers [13–15] (joint with H. Nakajima) where a mathematical definition of the Coulomb branches $\mathcal{M}_C(G, \mathbf{M})$ ia given under an additional assumption (namely, we assume that $\mathbf{M} = T^*\mathbf{N} = \mathbf{N} \oplus \mathbf{N}^*$ for some representation $\mathbf{N}$ of G—such theories are called *gauge theories of cotangent type*) and some further properties of Coulomb brancnes are studied.[3] In this case we shall write $\mathcal{M}_C(G, \mathbf{N})$ instead of $\mathcal{M}_C(G, \mathbf{M})$. In *loc. cit.* it is defined as $\mathrm{Spec}(\mathbb{C}[\mathcal{M}_C(G, \mathbf{N})])$ where $\mathbb{C}[\mathcal{M}_C(G, \mathbf{N})]$ is some geometrically defined algebra over $\mathbb{C}$. The varieties $\mathcal{M}_C(G, \mathbf{N})$ are normal, affine, Poisson and generically symplectic. We expect that the they are actually singular symplectic, but we can't prove this in general. The main ingredient in the definition is the geometry of the affine Grassmannian Gr_G of G.

1.1.9 Remark About Categorical Symplectic Duality

The following will never be used in the sequel, but we think it is important to mention it for the interested reader. Perhaps the most interesting aspect of symplectic duality is the *categorical symplectic duality* discussed in [10]. Namely, in *loc. cit.* the authors explain that if both X and X^* have a symplectic resolution, then one can think about symplectic duality between them as Koszul duality between some version of category $\mathcal{O}$ over the quantization of the algebras $\mathbb{C}[X]$ and $\mathbb{C}[X^*]$. In fact, it is explained in [47] that a slightly weaker version of this statement can be formulated even when X and X^* do not have a symplectic resolution. This weaker statement is in fact proved in [47] for $\mathcal{M}_H(G, \mathbf{N})$ and $\mathcal{M}_C(G, \mathbf{N})$ (where the author uses the definition of $\mathcal{M}_C(G, \mathbf{N})$ from [13]).

[2] $(T^*T^\vee)/W$ is actually the Coulomb branch of the corresponding classical field theory and the fact that the above birational isomorphism is not in general biregular means that "in the quantum theory the Coulomb branch acquires quantum corrections".

[3] The reader is also advised to consult the papers [39, 41, 42] by Nakajima. In particular, the papers [13–15] are based on the ideas developed earlier in [39]. Also [41] contains a lot of interesting open problems in the subject most of which will not be addressed in these notes.

1.1.10 The Plan

These notes are organized as follows. First, as was mentioned above the main geometric player in our construction of $\mathcal{M}_C(G,\mathbf{N})$ is the affine Grassmannian Gr_G of G. In Sects. 1.2 and 1.3 we review some facts and constructions related to Gr_G. Namely, in Sect. 1.2 we review the so called *geometric Satake equivalence*; in Sect. 1.3 we discuss an upgrade this construction: the *derived geometric Satake equivalence*. In Sect. 1.4 we discuss some general expectations about 3d $N = 4$ theories and in Sect. 1.5 we give a definition of the varieties $\mathcal{M}_C(G,\mathbf{N})$. Section 1.6 is devoted to the example of quiver gauge theories; in particular, for finite quivers of ADE type we identify the Coulomb branches with certain (generalized) slices in the affine Grassmannian of the corresponding group of ADE-type. Finally, in Sect. 1.7 we discuss some conjectural categorical structures related to the *topologically twisted version* of gauge theories of cotangent type (we have learned the main ideas of this section from T. Dimofte, D. Gaiotto, J. Hilburn and P. Yoo).

1.2 Geometric Satake

1.2.1 Overview

Let $\mathcal{O}$ denote the formal power series ring $\mathbb{C}[[z]]$, and let $\mathcal{K}$ denote its fraction field $\mathbb{C}((z))$. Let G be a reductive complex algebraic group with a Borel and a Cartan subgroup $G \supset B \supset T$, and with the Weyl group W of (G,T). Let Λ be the coweight lattice, and let $\Lambda^+ \subset \Lambda$ be the submonoid of dominant coweights. Let also $\Lambda_+ \subset \Lambda$ be the submonoid spanned by the simple coroots α_i, $i \in I$. We denote by $G^\vee \supset T^\vee$ the Langlands dual group, so that Λ is the weight lattice of $G^\vee$.

The affine Grassmannian $\mathrm{Gr}_G = G_\mathcal{K}/G_\mathcal{O}$ is an ind-projective scheme, the union $\bigsqcup_{\lambda\in\Lambda^+}\mathrm{Gr}_G^\lambda$ of $G_\mathcal{O}$-orbits. The closure of Gr_G^λ is a projective variety $\overline{\mathrm{Gr}}_G^\lambda = \bigsqcup_{\mu\leq\lambda}\mathrm{Gr}_G^\mu$. The fixed point set Gr_G^T is naturally identified with the coweight lattice Λ; and $\mu \in \Lambda$ lies in $\overline{\mathrm{Gr}}_G^\lambda$ iff $\mu \in W\lambda$.

One of the cornerstones of the Geometric Langlands Program initiated by V. Drinfeld is an equivalence $\mathbb{S}$ of the tensor category $\mathrm{Rep}(G^\vee)$ and the category $\mathrm{Perv}_{G_\mathcal{O}}(\mathrm{Gr}_G)$ of $G_\mathcal{O}$-equivariant perverse constructible sheaves on Gr_G equipped with a natural monoidal convolution structure $\star$ and a fiber functor $H^\bullet(\mathrm{Gr}_G, -)$ [5, 25, 36, 37]. It is a categorification of the classical Satake isomorphism between $K(\mathrm{Rep}(G^\vee)) = \mathbb{C}[T^\vee]^W$ and the spherical affine Hecke algebra of G. The geometric Satake equivalence $\mathbb{S}$ sends an irreducible $G^\vee$-module V^λ with highest weight λ to the Goresky–MacPherson sheaf $\mathrm{IC}(\overline{\mathrm{Gr}}_G^\lambda)$.

In order to construct a commutativity constraint for $(\mathrm{Perv}_{G_\mathcal{O}}(\mathrm{Gr}_G), \star)$, Beilinson and Drinfeld introduced a relative version $\mathrm{Gr}_{G,BD}$ of the Grassmannian over the Ran space of a smooth curve X, and a fusion monoidal structure Ψ on $\mathrm{Perv}_{G_\mathcal{O}}(\mathrm{Gr}_G)$ (isomorphic to $\star$). One of the main discoveries of [37] was a Λ-grading of the fiber functor $H^\bullet(\mathrm{Gr}_G, \mathcal{F}) = \bigoplus_{\lambda\in\Lambda} \Phi_\lambda(\mathcal{F})$ by the hyperbolic stalks at T-fixed points. For a $G^\vee$-module V, its weight space V_λ is canonically isomorphic to the hyperbolic stalk $\Phi_\lambda(\mathbb{S}V)$.

Various geometric structures of a perverse sheaf $\mathbb{S}V$ reflect some fine representation theoretic structures of V, such as Brylinski–Kostant filtration and the action of dynamical Weyl group, see [28]. One of the important technical tools of studying $\mathrm{Perv}_{G_\mathcal{O}}(\mathrm{Gr}_G)$ is the embedding $\mathrm{Gr}_G \hookrightarrow \mathbf{Gr}_G$ into Kashiwara infinite type scheme $\mathbf{Gr}_G = G_{\mathbb{C}((z^{-1}))}/G_{\mathbb{C}[z]}$ [30, 31]. The quotient $G_{\mathbb{C}[[z^{-1}]]}\backslash\mathbf{Gr}_G$ is the moduli stack $\mathrm{Bun}_G(\mathbb{P}^1)$ of G-bundles on the projective line $\mathbb{P}^1$. The $G_{\mathbb{C}[[z^{-1}]]}$-orbits on $\mathbf{Gr}_G$ are of finite codimension; they are also numbered by the dominant coweights of G, and the image of an orbit $\mathbf{Gr}^\lambda_G$ in $\mathrm{Bun}_G(\mathbb{P}^1)$ consists of G-bundles of isomorphism type λ [29]. The stratifications $\mathrm{Gr}_G = \bigsqcup_{\lambda\in\Lambda^+} \mathrm{Gr}^\lambda_G$ and $\mathbf{Gr}_G = \bigsqcup_{\lambda\in\Lambda^+} \mathbf{Gr}^\lambda_G$ are transversal, and their intersections and various generalizations thereof will play an important role later on.

More precisely, we denote by K_1 the first congruence subgroup of $G_{\mathbb{C}[[z^{-1}]]}$: the kernel of the evaluation projection $\mathrm{ev}_\infty\colon G_{\mathbb{C}[[z^{-1}]]} \twoheadrightarrow G$. The transversal slice $\mathcal{W}^\lambda_\mu$ (resp. $\overline{\mathcal{W}}^\lambda_\mu$) is defined as the intersection of Gr^λ_G (resp. $\overline{\mathrm{Gr}}^\lambda_G$) and $K_1 \cdot \mu$ in $\mathbf{Gr}_G$. It is known that $\overline{\mathcal{W}}^\lambda_\mu$ is nonempty iff $\mu \leq \lambda$, and $\dim \overline{\mathcal{W}}^\lambda_\mu$ is an affine irreducible variety of dimension $\langle 2\rho^\vee, \lambda - \mu\rangle$. Following an idea of Mirković, [32] proved that $\overline{\mathcal{W}}^\lambda_\mu = \bigsqcup_{\mu\leq\nu\leq\lambda} \mathcal{W}^\nu_\mu$ is the decomposition of $\overline{\mathcal{W}}^\lambda_\mu$ into symplectic leaves of a natural Poisson structure.

1.2.2 *Hyperbolic Stalks*

Let N denote the unipotent radical of the Borel B, and let N_- stand for the unipotent radical of the opposite Borel B_-. For a coweight $\nu \in \Lambda = \mathrm{Gr}^T_G$, we denote by $S_\nu \subset \mathrm{Gr}_G$ (resp. $T_\nu \subset \mathrm{Gr}_G$) the orbit of $N(\mathcal{K})$ (resp. of $N_-(\mathcal{K})$). The intersections $S_\nu \cap \overline{\mathrm{Gr}}^\lambda_G$ (resp. $T_\nu \cap \overline{\mathrm{Gr}}^\lambda_G$) are the *attractors* (resp. *repellents*) of $\mathbb{C}^\times$ acting via its homomorphism 2ρ to the Cartan torus $T \curvearrowright \overline{\mathrm{Gr}}^\lambda_G$: $S_\nu \cap \overline{\mathrm{Gr}}^\lambda_G = \{x \in \overline{\mathrm{Gr}}^\lambda_G : \lim_{c\to 0} 2\rho(c)\cdot x = \nu\}$ and $T_\nu \cap \overline{\mathrm{Gr}}^\lambda_G = \{x \in \overline{\mathrm{Gr}}^\lambda_G : \lim_{c\to\infty} 2\rho(c)\cdot x = \nu\}$. Going to the limit $\mathrm{Gr}_G = \lim_{\lambda\in\Lambda^+} \overline{\mathrm{Gr}}^\lambda_G$, S_ν (resp. T_ν) is the attractor (resp. repellent) of ν in Gr_G. We denote by $r_{\nu,+}$ (resp. $r_{\nu,-}$) the locally closed embedding $S_\nu \hookrightarrow \mathrm{Gr}_G$ (resp. $T_\nu \hookrightarrow \mathrm{Gr}_G$). We also denote by $\iota_{\nu,+}$ (resp. $\iota_{\nu,-}$) the closed embedding of the point ν into S_ν (resp. into T_ν). The following theorem is proved in [8, 18].

Theorem 1.1 *There is a canonical isomorphism of functors* $\iota^*_{\nu,+}r^!_{\nu,+} \simeq \iota^!_{\nu,-}r^*_{\nu,-}\colon D^b_{G_{\mathcal{O}}}(\mathrm{Gr}_G) \to D^b(\mathrm{Vect})$.

Definition 1.2 For a sheaf $\mathcal{F} \in D^b_{G_{\mathcal{O}}}(\mathrm{Gr}_G)$ we define its hyperbolic stalk at ν as $\Phi_\nu(\mathcal{F}) := \iota^*_{\nu,+}r^!_{\nu,+}\mathcal{F} \simeq \iota^!_{\nu,-}r^*_{\nu,-}\mathcal{F}$.

The following dimension estimate due to [37] is crucial for the geometric Satake.

Lemma 1.3

(a) $S_\nu \cap \mathrm{Gr}^\lambda_G \neq \varnothing$ *iff* $T_\nu \cap \mathrm{Gr}^\lambda_G \neq \varnothing$ *iff* ν *has nonzero multiplicity in the irreducible* $G^\vee$*-module* V^λ *with highest weight* λ. *This is also equivalent to* $\nu \in \overline{\mathrm{Gr}}^\lambda_G$.
(b) *The nonempty intersection* $S_\nu \cap \overline{\mathrm{Gr}}^\lambda_G$ *is equidimensional of dimension* $\langle \nu+\lambda, \rho^\vee\rangle$.
(c) *The nonempty intersection* $T_\nu \cap \overline{\mathrm{Gr}}^\lambda_G$ *is equidimensional of dimension* $\langle \nu + w_0\lambda, \rho^\vee\rangle$. *Here* w_0 *is the longest element of the Weyl group* W.

Corollary 1.4

(a) *For* $\mathcal{F} \in \mathrm{Perv}_{G_{\mathcal{O}}}(\mathrm{Gr}_G)$, *the hyperbolic stalk* $\Phi_\nu(\mathcal{F})$ *is concentrated in degree* $\langle \nu, 2\rho^\vee\rangle$.
(b) *There is a canonical direct sum decomposition* $H^\bullet(\mathrm{Gr}_G, \mathcal{F}) = \bigoplus_{\nu\in\Lambda} \Phi_\nu(\mathcal{F})$.
(c) *The functor* $H^\bullet(\mathrm{Gr}_G, -)\colon \mathrm{Perv}_{G_{\mathcal{O}}}(\mathrm{Gr}_G) \to \mathrm{Vect}^{\mathrm{gr}}$, *as well as its upgrade* $\bigoplus_{\nu\in\Lambda}\Phi_\nu\colon \mathrm{Perv}_{G_{\mathcal{O}}}(\mathrm{Gr}_G) \to \mathrm{Rep}(T^\vee)$, *is exact and conservative.*

1.2.3 Convolution

We have the following basic diagram:

$$\mathrm{Gr}_G \times \mathrm{Gr}_G \xleftarrow{p} G_{\mathcal{K}} \times \mathrm{Gr}_G \xrightarrow{q} \mathrm{Gr}_G \tilde{\times} \mathrm{Gr}_G \xrightarrow{m} \mathrm{Gr}_G. \tag{1.1}$$

Here $\mathrm{Gr}_G \tilde{\times} \mathrm{Gr}_G = G_{\mathcal{K}} \overset{G_{\mathcal{O}}}{\times} \mathrm{Gr}_G = (G_{\mathcal{K}} \times \mathrm{Gr}_G)/((g,x) \sim (gh^{-1}, hx),\ h \in G_{\mathcal{O}})$. Furthermore, p is the projection on the first factor and identity on the second factor, and the composition $m \circ q$ is the action morphism $G_{\mathcal{K}} \times \mathrm{Gr}_G \to \mathrm{Gr}_G$ (which clearly factors through $G_{\mathcal{K}} \overset{G_{\mathcal{O}}}{\times} \mathrm{Gr}_G$).

Definition 1.5 Given $\mathcal{F}_1, \mathcal{F}_2 \in D^b_{G_{\mathcal{O}}}(\mathrm{Gr}_G)$, their convolution $\mathcal{F}_1 \star \mathcal{F}_2 \in D^b_{G_{\mathcal{O}}}(\mathrm{Gr}_G)$ is defined as $\mathcal{F}_1 \star \mathcal{F}_2 := m_*(\mathcal{F}_1 \tilde{\boxtimes} \mathcal{F}_2)$, where $\mathcal{F}_1 \tilde{\boxtimes} \mathcal{F}_2$ is the descent of $p^*(\mathcal{F}_1 \boxtimes \mathcal{F}_2)$, that is $q^*(\mathcal{F}_1 \tilde{\boxtimes} \mathcal{F}) = p^*(\mathcal{F}_1 \boxtimes \mathcal{F}_2)$.

The next lemma is due to [36, 37]. It follows from the *stratified semismallness* of $m\colon \overline{\mathrm{Gr}}^{\lambda,\mu}_G := p^{-1}(\overline{\mathrm{Gr}}^\lambda_G) \overset{G_{\mathcal{O}}}{\times} \overline{\mathrm{Gr}}^\mu_G \to \overline{\mathrm{Gr}}^{\lambda+\mu}_G$. Here $\overline{\mathrm{Gr}}^{\lambda,\mu}_G$ is stratified by the union of $\mathrm{Gr}_G^{\nu,\theta} := p^{-1}(\mathrm{Gr}_G^{\nu}) \overset{G_\mathcal{O}}{\times} \mathrm{Gr}_G^{\theta}$ over $\nu \leq \lambda,\ \theta \leq \mu$. The stratified semismallness in turn follows from the dimension estimate of Lemma 1.3.

Lemma 1.6 *Given* $\mathcal{F}_1, \mathcal{F}_2 \in \mathrm{Perv}_{G_\mathcal{O}}(\mathrm{Gr}_G)$, *their convolution* $\mathcal{F}_1 \star \mathcal{F}_2$ *lies in* $\mathrm{Perv}_{G_\mathcal{O}}(\mathrm{Gr}_G)$ *as well.*

In order to define a commutativity constraint for $\star$, we will need an equivalent construction of the monoidal structure on $\mathrm{Perv}_{G_\mathcal{O}}(\mathrm{Gr}_G)$ via *fusion* due to V. Drinfeld.

1.2.4 Fusion

Let X be a smooth curve, e.g. $X = \mathbb{A}^1$. We have the following basic diagram:

$$\begin{array}{ccccc}
(\mathrm{Gr}_G \tilde{\times} \mathrm{Gr}_G)_X & \xrightarrow{i} & \mathrm{Gr}_{G,X} \tilde{\times} \mathrm{Gr}_{G,X} & \xleftarrow{j} & (\mathrm{Gr}_{G,X} \times \mathrm{Gr}_{G,X})|_U \\
{\scriptstyle m_X}\downarrow & & {\scriptstyle m_{X^2}}\downarrow & & \wr\downarrow \\
\mathrm{Gr}_{G,X} & \xrightarrow{i} & \mathrm{Gr}_{G,X^2} & \xleftarrow{j} & (\mathrm{Gr}_{G,X} \times \mathrm{Gr}_{G,X})|_U \\
{\scriptstyle \pi}\downarrow & & {\scriptstyle \pi}\downarrow & & {\scriptstyle \pi}\downarrow \\
X & \xrightarrow{\Delta} & X^2 & \xleftarrow{j} & U.
\end{array}$$

Here $U \hookrightarrow X^2$ is the open embedding of the complement to the diagonal $\Delta_X \hookrightarrow X^2$. Furthermore, for $n \in \mathbb{N}$, Gr_{G,X^n} is the moduli space of the following data: $\{(x_1, \ldots, x_n) \in X^n,\ \mathcal{P}_G,\ \tau\}$, where $\mathcal{P}_G$ is a G-bundle on X, and τ is a trivialization of $\mathcal{P}_G$ on $X \setminus \{x_1, \ldots, x_n\}$. The projection $\pi : \mathrm{Gr}_{G,X^n} \to X^n$ forgets the data of $\mathcal{P}_G$ and τ. Note that $\mathrm{Gr}_{G,X^2}|_U \simeq (\mathrm{Gr}_{G,X} \times \mathrm{Gr}_{G,X})|_U$, while $\mathrm{Gr}_{G,X^2}|_{\Delta_X} \simeq \mathrm{Gr}_{G,X}$. Furthermore, $\mathrm{Gr}_{G,X} \tilde{\times} \mathrm{Gr}_{G,X}$ is the moduli space of the following data: $\{(x_1, x_2) \in X^2,\ \mathcal{P}_G^1, \mathcal{P}_G^2, \tau, \sigma\}$, where $\mathcal{P}_G^1, \mathcal{P}_G^2$ are G-bundles on X; $\sigma : \mathcal{P}_G^1|_{X \setminus x_2} \xrightarrow{\sim} \mathcal{P}_G^2|_{X \setminus x_2}$, and τ is a trivialization of $\mathcal{P}_G^1$ on $X \setminus x_1$. Note that $(\mathrm{Gr}_{G,X} \tilde{\times} \mathrm{Gr}_{G,X})|_U \simeq (\mathrm{Gr}_{G,X} \times \mathrm{Gr}_{G,X})|_U$, while $(\mathrm{Gr}_{G,X} \tilde{\times} \mathrm{Gr}_{G,X})|_{\Delta_X}$ is fibered over X with fibers isomophic to $\mathrm{Gr}_G \tilde{\times} \mathrm{Gr}_G$. Finally, $m_{X^2} : \mathrm{Gr}_{G,X} \tilde{\times} \mathrm{Gr}_{G,X} \to \mathrm{Gr}_{G,X^2}$ takes $(x_1, x_2, \mathcal{P}_G^1, \mathcal{P}_G^2, \tau, \sigma)$ to $(x_1, x_2, \mathcal{P}_G^2, \tau')$ where $\tau' = \sigma \circ \tau|_{X \setminus \{x_1, x_2\}}$. All the squares in the above diagram are cartesian. The stratified semismallness property of the convolution morphism m used in the proof of Lemma 1.6 implies the *stratified smallness* property of the relative convolution morphism m_{X^2}.

Now given $\mathcal{F}_1, \mathcal{F}_2 \in D^b_{G_\mathcal{O}}(\mathrm{Gr}_G)$, we can define the constructible complexes $\mathcal{F}_{1,X}, \mathcal{F}_{2,X}$ on $\mathrm{Gr}_{G,X}$ smooth over X, and by descent similarly to Sect. 1.2.3, a constructible complex $\mathcal{F}_{1,X} \tilde{\boxtimes} \mathcal{F}_{2,X}$ on $\mathrm{Gr}_{G,X} \tilde{\times} \mathrm{Gr}_{G,X}$ smooth over X^2. Note that $(\mathcal{F}_{1,X} \tilde{\boxtimes} \mathcal{F}_{2,X})|_U = (\mathcal{F}_{1,X} \boxtimes \mathcal{F}_{2,X})|_U$. For simplicity, let us take $X = \mathbb{A}^1$. Then by the proper base change for nearby cycles $\Psi_{x_1 - x_2} m_{X^2 *}(\mathcal{F}_{1,X} \tilde{\boxtimes} \mathcal{F}_{2,X})$ on Gr_{G,X^2} we deduce $(\mathcal{F}_1 \star \mathcal{F}_2)_X = \Psi_{x_1 - x_2}(\mathcal{F}_{1,X} \boxtimes \mathcal{F}_{2,X})|_U$. The RHS being manifestly symmetric, we deduce the desired commutativity constraint for the convolution product $\star$. Note that due to the stratified smallness of m_{X^2} and the local acyclicity of $\pi \circ m_{X^2}$, we have an isomorphism

$$\Psi_{x_1-x_2}(\mathcal{F}_{1,X} \boxtimes \mathcal{F}_{2,X})|_U[1] \xrightarrow{\sim} i^*(\pi \circ m_{X^2})_*(\mathcal{F}_{1,X} \boxtimes \mathcal{F}_{2,X})$$
$$\xrightarrow{\sim} i^* j_{!*}\left((\mathcal{F}_{1,X} \boxtimes \mathcal{F}_{2,X})|_U\right).$$

Also, the above smoothness of $\mathcal{F}_{1,X}\widetilde{\boxtimes}\mathcal{F}_{2,X}$ over X^2 implies that $\pi_* m_{X^2*}$ $(\mathcal{F}_{1,X}\widetilde{\boxtimes}\mathcal{F}_{2,X})$ is a constant sheaf on X^2. Since its diagonal stalks are $H^\bullet(\mathrm{Gr}_G, \mathcal{F}_1 \star \mathcal{F}_2)$, and the off-diagonal stalks are $H^\bullet(\mathrm{Gr}_G, \mathcal{F}_1) \otimes H^\bullet(\mathrm{Gr}_G, \mathcal{F}_2)$, we obtain that the cohomology functor $\mathrm{Perv}_{G_\mathcal{O}}(\mathrm{Gr}_G) \to \mathrm{Vect}$ is a *tensor* functor: $H^\bullet(\mathrm{Gr}_G, \mathcal{F}_1 \star \mathcal{F}_2) \xrightarrow{\sim} H^\bullet(\mathrm{Gr}_G, \mathcal{F}_1) \otimes H^\bullet(\mathrm{Gr}_G, \mathcal{F}_2)$.

Corollary 1.7 *The abelian category* $\mathrm{Perv}_{G_\mathcal{O}}(\mathrm{Gr}_G)$ *is equipped with a symmetric monoidal structure* $\star$ *and a fiber functor* $H^\bullet(\mathrm{Gr}_G, -)$.

By Tannakian formalism, the tensor category $\mathrm{Perv}_{G_\mathcal{O}}(\mathrm{Gr}_G)$ must be equivalent to $\mathrm{Rep}(G')$ for a proalgebraic group G'. It remains to identify G' with the Langlands dual group $G^\vee$. From the semisimplicity of $\mathrm{Perv}_{G_\mathcal{O}}(\mathrm{Gr}_G)$, the group G' must be reductive. The upgraded fiber functor $\bigoplus_{\nu\in\Lambda} \Phi_\nu\colon \mathrm{Perv}_{G_\mathcal{O}}(\mathrm{Gr}_G) \to \mathrm{Rep}(T^\vee)$ is tensor since the nearby cycles commute with hyperbolic stalks by [40, Proposition 5.4.1.(2)]. Hence we obtain a homomorphism $T^\vee \hookrightarrow G'$. Now it is easy to identify G' with $G^\vee$ using Lemma 1.3(a). We have proved

Theorem 1.8 *There is a tensor equivalence* $\mathbb{S}\colon \mathrm{Rep}(G^\vee), \otimes \xrightarrow{\sim} \mathrm{Perv}_{G_\mathcal{O}}(\mathrm{Gr}_G), \star$.

1.3 Derived Geometric Satake

In this section we extend the algebraic description of $\mathrm{Perv}_{G_\mathcal{O}}(\mathrm{Gr}_G)$ to an algebraic description of the equivariant derived category $D_{G_\mathcal{O}\rtimes\mathbb{C}^\times}(\mathrm{Gr}_G)$. Our exposition follows [6].

1.3.1 Asymptotic Harish-Chandra Bimodules

First we develop the necessary algebraic machinery. Let $U = U(\mathfrak{g}^\vee)$ be the universal enveloping algebra, and let $U_\hbar$ be the graded enveloping algebra, i.e. the graded $\mathbb{C}[\hbar]$-algebra generated by $\mathfrak{g}^\vee$ with relations $xy - yx = \hbar[x, y]$ for $x, y \in \mathfrak{g}^\vee$ (thus $U_\hbar$ is obtained from U by the Rees construction producing a graded algebra from the filtered one). The adjoint action extends to the action of $G^\vee$ on $U_\hbar$.

We consider the category $\mathcal{HC}_\hbar$ of graded modules over $U_\hbar^2 := U_\hbar \otimes_{\mathbb{C}[\hbar]} U_\hbar \simeq U_\hbar \rtimes U$ equipped with an action of $G^\vee$ (denoted $\beta\colon G^\vee \times M \to M$) satisfying the following conditions:

(a) The action $U_\hbar^2 \otimes M \to M$ is $G^\vee$-equivariant;
(b) for any $x \in \mathfrak{g}^\vee$, the action of $x \otimes 1 + 1 \otimes x \in U_\hbar^2$ coincides with the action of $\hbar \cdot d\beta(x)$;
(c) the module M is finitely generated as a $U_\hbar \otimes 1$-module (equivalently, as a $1 \otimes U_\hbar$-module).

The restriction from $U_\hbar^2$ to $U_\hbar \otimes 1$ gives an equivalence of $\mathcal{HC}_\hbar$ with the category of $G^\vee$-modules equipped with a $G^\vee$-equivariant $U_\hbar$-action.

1.3.1.1 Example: Free Harish-Chandra Bimodules

Let $V \in \operatorname{Rep}(G^\vee)$. We define a free Harish-Chandra bimodule $\operatorname{Fr}(V) = U_\hbar \otimes V$ with the $G^\vee$-action $g(y \otimes v) = \operatorname{Ad}_g(y) \otimes g(v)$ and the $U_\hbar^2$-action $(x \otimes u)(y \otimes v) = xyu \otimes v + \hbar xy \otimes u(v)$, where $x, u \in \mathfrak{g}^\vee \subset U_\hbar$. Thus, $\operatorname{Fr}(V)$ is the induction of V (the left adjoint functor to the restriction $\operatorname{res}\colon \mathcal{HC}_\hbar \to \operatorname{Rep}(G^\vee)$). This is a projective object of the category $\mathcal{HC}_\hbar$. We will denote by $\mathcal{HC}_\hbar^{\mathrm{fr}}$ the full subcategory of $\mathcal{HC}_\hbar$ formed by all the free Harish-Chandra bimodules.

1.3.2 Kostant–Whittaker Reduction

We also consider the subalgebra $U_\hbar^2(\mathfrak{n}_-^\vee) = U_\hbar(\mathfrak{n}_-^\vee) \rtimes U(\mathfrak{n}_-^\vee) \subset U_\hbar^2$. We fix a regular character $\psi\colon U_\hbar(\mathfrak{n}_-^\vee) \to \mathbb{C}[\hbar]$ taking value 1 at each generator f_i. We extend it to a character $\psi^{(2)}\colon U_\hbar^2(\mathfrak{n}_-^\vee) = U_\hbar(\mathfrak{n}_-^\vee) \rtimes U(\mathfrak{n}_-^\vee) \to \mathbb{C}[\hbar]$ trivial on the second factor (its restriction to $1 \otimes U_\hbar(\mathfrak{n}_-^\vee)$ equals $-\psi$).

Definition 1.9 For $M \in \mathcal{HC}_\hbar$ we set $\varkappa_\hbar(M) := (M \overset{L}{\otimes}_{1\otimes U_\hbar(\mathfrak{n}_-^\vee)} (-\psi))^{N_-^\vee}$ *(Kostant-Whittaker reduction)*. It is equipped with an action of the Harish-Chandra center $Z(U_\hbar) \otimes_{\mathbb{C}[\hbar]} Z(U_\hbar) = \mathbb{C}[\mathfrak{t}/W \times \mathfrak{t}/W \times \mathbb{A}^1]$. Furthermore, $\varkappa_\hbar(M)$ is graded by the action of the element h from a principal $\mathfrak{sl}_2 = \langle e, h, f\rangle$-triple whose e corresponds to ψ under the Killing form. All in all, $\varkappa_\hbar(M)$ is a $\mathbb{C}^\times$-equivariant coherent sheaf on $\mathfrak{t}/W \times \mathfrak{t}/W \times \mathbb{A}^1$ (with respect to the natural $\mathbb{C}^\times$-action on $\mathfrak{t}/W$).

1.3.2.1 Example: Differential Operators and Quantum Toda Lattice

The ring of $\hbar$-differential operators on $G^\vee$, $\mathcal{D}_\hbar(G^\vee) = U_\hbar \ltimes \mathbb{C}[G^\vee]$ is an object of the Ind-completion $\operatorname{Ind}\mathcal{HC}_\hbar$. It carries one more commuting structure of a Harish-Chandra bimodule (where $U_\hbar$ acts by the right-invariant $\hbar$-differential operators). We define $\mathcal{K} := \varkappa_\hbar(\mathcal{D}_\hbar(G^\vee))$, an algebra in the category $\operatorname{Ind}\mathcal{HC}_\hbar$. It corepresents the functor $\mathrm{Hom}(M, \mathcal{K}) = \varkappa_\hbar(DM)$ where $DM = \mathrm{Hom}_{U_\hbar}(M, U_\hbar)$ is a duality on $\mathcal{HC}_\hbar$. Here $\mathrm{Hom}_{U_\hbar}$ is taken with respect to the *right* action of $U_\hbar$, while the *left* actions of $U_\hbar$ on M and on itself are used to construct the left and right actions of $U_\hbar$ on $\mathrm{Hom}_{U_\hbar}(M, U_\hbar)$. Finally, we define an associative algebra $\mathcal{T}_\hbar := \varkappa_\hbar(\mathcal{K}) = \varkappa_\hbar^{\mathrm{right}} \varkappa_\hbar^{\mathrm{left}} \mathcal{D}_\hbar(G^\vee)$, the quantum open Toda lattice.

1.3.3 Deformation to the Normal Cone

A well known construction associates to a closed subvariety $Z \subset X$ the deformation to the normal cone $\mathcal{N}_Z X$ projecting to $X \times \mathbb{A}^1$; all the nonzero fibers are isomorphic to X, while the zero fiber is isomorphic to the normal cone $C_Z X$. Recall that $\mathcal{N}_Z X$ is defined as the relative spectrum of the subsheaf of algebras $\mathcal{O}_X[\hbar^{\pm 1}]$, generated over $\mathcal{O}_{X\times\mathbb{A}^1}$ by the elements of the following type: $f\hbar^{-1}$, where $f \in \mathcal{O}_X,\ f|_Z = 0$.

Now if M is a Harish-Chandra bimodule free over $\mathbb{C}[\hbar]$, then the action of $\mathbb{C}[\mathfrak{t}/W \times \mathfrak{t}/W \times \mathbb{A}^1]$ on $\varkappa_\hbar(M)$ extends uniquely to the action of the ring of functions $\mathbb{C}[\mathcal{N}_\Delta(\mathfrak{t}/W \times \mathfrak{t}/W)]$ on the deformation to the normal cone of diagonal, since for $z \in ZU_\hbar,\ m \in M$, the difference of the left and right actions $z^{(1)}m - z^{(2)}m$ is divisible by $\hbar$. So we will consider $\varkappa_\hbar(M)$ as a $\mathbb{C}^\times$-equivariant sheaf on $\mathcal{N}_\Delta(\mathfrak{t}/W \times \mathfrak{t}/W)$. Note that $\mathbb{C}[\mathcal{N}_\Delta(\mathfrak{t}/W \times \mathfrak{t}/W)]/\hbar = \mathbb{C}[C_\Delta(\mathfrak{t}/W \times \mathfrak{t}/W)] = \mathbb{C}[T_{\mathfrak{t}/W}] = \mathbb{C}[T_\Sigma]$. Here $\Sigma \subset (\mathfrak{g}^\vee)^*$ is the Kostant slice (we identify $\mathfrak{g}^\vee$ and $(\mathfrak{g}^\vee)^*$ by the Killing form). Recall the universal centralizer $\mathfrak{Z}_{\mathfrak{g}^\vee}^{\mathfrak{g}^\vee} = \{(x \in \mathfrak{g}^\vee,\ \xi \in \Sigma) : \mathrm{ad}_x \xi = 0\}$.

Lemma 1.10 *Under the identification of $\mathfrak{g}^\vee$ and $(\mathfrak{g}^\vee)^*$, the universal centralizer $\mathfrak{Z}_{\mathfrak{g}^\vee}^{\mathfrak{g}^\vee}$ is canonically isomorphic to the cotangent bundle $T^*\Sigma$.*

Proof The open subset of regular elements $(\mathfrak{g}^\vee)^*_{\mathrm{reg}} \subset (\mathfrak{g}^\vee)^*$ carries the centralizer sheaf $\mathfrak{z} \subset \mathfrak{g}^\vee \otimes \mathcal{O}$ of abelian Lie algebras. We have $\mathfrak{z} = \mathrm{pr}^* T^*\Sigma$ where $\mathrm{pr}\colon (\mathfrak{g}^\vee)^*_{\mathrm{reg}} \to (\mathfrak{g}^\vee)^*_{\mathrm{reg}}/\mathrm{Ad}_{G^\vee} = \mathfrak{t}/W = \Sigma$ is the evident projection. Indeed, the fiber of $\mathrm{pr}^* T^*\Sigma$ at a point $\eta \in (\mathfrak{g}^\vee)^*$ is dual to the cokernel of the map $a_\eta\colon \mathfrak{g}^\vee \to (\mathfrak{g}^\vee)^*,\ x \mapsto \mathrm{ad}_x \eta$. In other words, this fiber is isomorphic to the kernel of the dual map which happens to coincide with a_η. The latter kernel is by definition nothing but the fiber $\mathfrak{z}_\eta$. □

Lemma 1.11 *For $V \in \mathrm{Rep}(G^\vee)$, the $\mathbb{C}[T_\Sigma]$-module $\varkappa_\hbar(\mathrm{Fr}(V))|_{\hbar=0}$ is isomorphic to $\mathbb{C}[\Sigma] \otimes V$ as a $\mathfrak{Z}_{\mathfrak{g}^\vee}^{\mathfrak{g}^\vee}$-module.*

Proof Let $\mathrm{Poly}((\mathfrak{g}^\vee)^*, \mathfrak{g}^\vee)^{G^\vee}$ be the space of $G^\vee$-invariant polynomial morphisms. It is a vector bundle over $\mathrm{Spec}\,\mathbb{C}[(\mathfrak{g}^\vee)^*]^{G^\vee} = \Sigma$. If $P \in \mathbb{C}[(\mathfrak{g}^\vee)^*]^{G^\vee}$, then the differential dP is a section of this bundle. If $z \in ZU(\mathfrak{g}^\vee) = \mathbb{C}[(\mathfrak{g}^\vee)^*]^{G^\vee}$, we denote dz by σ_z. If $\{z_i\}$ is a set of generators of $ZU(\mathfrak{g}^\vee)$, then $\{\sigma_{z_i}\}$ forms a basis of $\mathfrak{Z}_{\mathfrak{g}^\vee}^{\mathfrak{g}^\vee}$ as a vector bundle over Σ, and hence identifies it with $T^*\Sigma$. Let $z^{(1)}, z^{(2)}$ stand for the left and right actions of z on $\mathrm{Fr}(V)$. We have to check that the action of $\frac{z^{(1)}-z^{(2)}}{\hbar}|_{\hbar=0}$ on $(U_\hbar \otimes V)|_{\hbar=0} = \mathbb{C}[(\mathfrak{g}^\vee)^*] \otimes V$ coincides with the action of $\sigma_z \in \mathbb{C}[(\mathfrak{g}^\vee)^*] \otimes \mathfrak{g}^\vee$.

But if $v \in V$, $z = \sum_{\underline{i}} c_{\underline{i}} x_{i_1} \cdots x_{i_r}$ $(x_{i_p} \in \mathfrak{g}^\vee)$, and $\tilde{y} \in U_\hbar$ is a lift of $y \in \mathbb{C}[(\mathfrak{g}^\vee)^*] = U_\hbar|_{\hbar=0}$, then $\frac{z(\tilde{y}\otimes v)-(\tilde{y}\otimes v)z}{\hbar}|_{\hbar=0} = \sum_{i_p \in \underline{i}} c_{\underline{i}} x_{i_1} \cdots \hat{x}_{i_p} \cdots x_{i_r} y \otimes x_{i_p}(v) = \sigma_z(y \otimes v)$. □

1.3.4 Quasiclassical Limit of the Kostant–Whittaker Reduction

We define a functor $\varkappa\colon \mathsf{Coh}^{G^\vee \times \mathbb{C}^\times}(\mathfrak{g}^\vee)^* \to \mathsf{Coh}^{\mathbb{C}^\times}(T\Sigma)$ as follows. For a $G^\vee$-equivariant coherent sheaf $\mathcal{F}$ on $(\mathfrak{g}^\vee)^*$, the restriction $\mathcal{F}|_{(\mathfrak{g}^\vee)^*_{\mathrm{reg}}}$ is equipped with an action of the centralizer sheaf $\mathfrak{z}$. Hence, this restriction gives rise to a coherent sheaf on $\mathrm{pr}^* T\Sigma$. Restricting it to the Kostant slice Σ, we obtain a coherent sheaf $\varkappa\mathcal{F}$ on $T\Sigma$. Equivalently, the latter sheaf can be described as $(\mathcal{F}|_\Upsilon)^{N^\vee_-}$, where $\Upsilon = e + \mathfrak{b}^\vee_- \subset \mathfrak{g}^\vee \simeq (\mathfrak{g}^\vee)^*$. This construction is $\mathbb{C}^\times$-equivariant, and gives rise to the desired functor $\varkappa$.

We define $\mathsf{Coh}^{G^\vee \times \mathbb{C}^\times}_{\mathrm{fr}}(\mathfrak{g}^\vee)^* \subset \mathsf{Coh}^{G^\vee \times \mathbb{C}^\times}(\mathfrak{g}^\vee)^*$ as the full subcategory formed by the sheaves $V \otimes \mathcal{O}_{(\mathfrak{g}^\vee)^*}$ for $V \in \mathrm{Rep}(G^\vee)$.

Lemma 1.12

(a) *The functors $\varkappa$, $\varkappa_\hbar$ are exact;*
(b) $\varkappa|_{\mathsf{Coh}^{G^\vee \times \mathbb{C}^\times}_{\mathrm{fr}}(\mathfrak{g}^\vee)^*}$, $\varkappa_\hbar|_{\mathcal{HC}^{\mathrm{fr}}_\hbar}$ *are fully faithful.*

Proof The statements about $\varkappa_\hbar$ follow from the ones for $\varkappa$ by the graded Nakayama Lemma. To prove (a), note that the functor $\mathcal{F} \mapsto \mathcal{F}|_\Upsilon$, $\mathsf{Coh}^{G^\vee}(\mathfrak{g}^\vee)^* \to \mathsf{Coh}^{N^\vee_-}(\Upsilon)$ is exact. Moreover, $N^\vee_-$ acts freely on Υ, and $\Upsilon/N^\vee_- = \Sigma$. It follows that the functor $\mathcal{G} \mapsto \mathcal{G}^{N^\vee_-}$ is exact on $\mathsf{Coh}^{N^\vee_-}(\Upsilon)$. Now (b) follows since the codimension in $(\mathfrak{g}^\vee)^*$ of the complement $(\mathfrak{g}^\vee)^* \setminus (\mathfrak{g}^\vee)^*_{\mathrm{reg}}$ is at least 2, and the centralizer of a general regular element is connected. □

1.3.5 Equivariant Cohomology of the Affine Grassmannian

Now we turn to the topological machinery. We have an evident homomorphism $\mathrm{pr}^*\colon \mathbb{C}[\hbar] = H^\bullet_{\mathbb{C}^\times}(\mathrm{pt}) \to H^\bullet_{G_\mathcal{O} \rtimes \mathbb{C}^\times}(\mathrm{Gr}_G)$. Also, viewing $H^\bullet_{G_\mathcal{O} \rtimes \mathbb{C}^\times}(\mathrm{Gr}_G)$ as $H^\bullet_{\mathbb{C}^\times}(G_\mathcal{O} \backslash G_\mathcal{K} / G_\mathcal{O})$, we obtain two homomorphisms $\mathrm{pr}_1^*, \mathrm{pr}_2^*\colon \mathbb{C}[\Sigma] = \mathbb{C}[\mathfrak{t}/W] = H^\bullet_{G_\mathcal{O}}(\mathrm{pt}) \rightrightarrows H^\bullet_{G_\mathcal{O} \rtimes \mathbb{C}^\times}(\mathrm{Gr}_G)$. Let us assume for simplicity that G is simply connected. Recall the deformation to the normal cone of diagonal in $\mathfrak{t}/W \times \mathfrak{t}/W$, see Sect. 1.3.3.

Proposition 1.13 *There is a natural isomorphism* $\alpha\colon \mathbb{C}[\mathbb{N}_\Delta(\mathfrak{t}/W \times \mathfrak{t}/W)] \xrightarrow{\sim} H^\bullet_{G_\mathcal{O} \rtimes \mathbb{C}^\times}(\mathrm{Gr}_G)$ *compatible with the above* $\mathrm{pr}_1^*, \mathrm{pr}_2^*, \mathrm{pr}^*$.

Proof Since $H^\bullet_{G_\mathcal{O} \rtimes \mathbb{C}^\times}(\mathrm{Gr}_G)|_{\hbar=0} = H^\bullet_{G_\mathcal{O}}(\mathrm{Gr}_G)$, we see that $\mathrm{pr}_1^*|_{\hbar=0} = \mathrm{pr}_2^*|_{\hbar=0}$. It follows that $(\mathrm{pr}_1^*, \mathrm{pr}_2^*, \mathrm{pr}^*)\colon \mathbb{C}[\mathfrak{t}/W \times \mathfrak{t}/W \times \mathbb{A}^1] \to H^\bullet_{G_\mathcal{O} \rtimes \mathbb{C}^\times}(\mathrm{Gr}_G)$ factors as a composition $\mathbb{C}[\mathfrak{t}/W \times \mathfrak{t}/W \times \mathbb{A}^1] \to \mathbb{C}[\mathcal{N}_\Delta(\mathfrak{t}/W \times \mathfrak{t}/W)] \xrightarrow{\alpha} H^\bullet_{G_\mathcal{O} \rtimes \mathbb{C}^\times}(\mathrm{Gr}_G)$ for a uniquely defined homomorphism α. Let us check that α is injective. Indeed, $\alpha_{\mathrm{loc}} \colon \mathbb{C}[\mathcal{N}_\Delta(\mathfrak{t}/W \times \mathfrak{t}/W)] \otimes_{\mathbb{C}[\mathfrak{t}/W \times \mathbb{A}^1]} \mathrm{Frac}(\mathbb{C}[\mathfrak{t} \times \mathbb{A}^1]) \to H^\bullet_{G_\mathcal{O} \rtimes \mathbb{C}^\times}(\mathrm{Gr}_G) \otimes_{\mathbb{C}[\mathfrak{t}/W \times \mathbb{A}^1]} \mathrm{Frac}(\mathbb{C}[\mathfrak{t} \times \mathbb{A}^1]) = H^\bullet_{T \times \mathbb{C}^\times}(\mathrm{Gr}_G) \otimes_{\mathbb{C}[\mathfrak{t} \times \mathbb{A}^1]} \mathrm{Frac}(\mathbb{C}[\mathfrak{t} \times \mathbb{A}^1]) \hookrightarrow \varprojlim H^\bullet_{T \times \mathbb{C}^\times}(\overline{\mathrm{Gr}}^\lambda_G) \otimes_{\mathbb{C}[\mathfrak{t} \times \mathbb{A}^1]} \mathrm{Frac}(\mathbb{C}[\mathfrak{t} \times \mathbb{A}^1]) = \prod_{\lambda \in X_*(T)} \mathrm{Frac}(\mathbb{C}[\mathfrak{t} \times \mathbb{A}^1])$ associates to a function its restriction to the union of graphs $\Gamma_\lambda := \{(x_1, x_2, c) : x_1 = x_2 + c\lambda\} \subset \mathfrak{t} \times \mathfrak{t} \times \mathbb{A}^1$. More precisely, for a T-fixed point $\lambda \in \mathrm{Gr}_G$, the $\mathbb{C}[\mathfrak{t} \times (\mathfrak{t}/W) \times \mathbb{A}^1]$-module $H^\bullet_{T \times \mathbb{C}^\times}(\lambda)$ is canonically isomorphic to $(\mathrm{Id}, \pi, \mathrm{Id})_* \mathcal{O}_{\Gamma_\lambda}$, where π stands for the projection $\mathfrak{t} \to \mathfrak{t}/W$. Indeed, let $p \colon \mathcal{F}\ell_G \to \mathrm{Gr}_G$ be the projection from the affine flag variety $\mathcal{F}\ell_G = G_\mathcal{K}/\mathrm{Iw}$ of G, and let $\tilde{\lambda} \in \mathcal{F}\ell_G$ be the T-fixed point in $p^{-1}(\lambda)$ corresponding to the coweight $\lambda \in X_*(T) \subset W_{\mathrm{aff}}$. Viewing $H^\bullet_{T \times \mathbb{C}^\times}(\mathcal{F}\ell_G)$ as $H^\bullet_{\mathbb{C}^\times}(\mathrm{Iw} \backslash G_\mathcal{K}/\mathrm{Iw})$, we identify $H^\bullet_{T \times \mathbb{C}^\times}(\tilde{\lambda})$ with a $\mathbb{C}[\mathfrak{t} \times \mathfrak{t} \times \mathbb{A}^1]$-module M such that $(\mathrm{Id}, \pi, \mathrm{Id})_* M = H^\bullet_{T \times \mathbb{C}^\times}(\lambda)$. The preimage T_λ of $\tilde{\lambda}$ in $G_\mathcal{K}$ is homotopy equivalent to the torus T, and the action of $T \times T \times \mathbb{C}^\times$ on T_λ is homotopy equivalent to $(t_1, t_2, c) \cdot t = t_1 t t_2^{-1} \lambda(c)$. We conclude that $H^\bullet_{T \times \mathbb{C}^\times}(\tilde{\lambda}) = H^\bullet_{\mathbb{C}^\times}(T \backslash T_\lambda / T) = \mathbb{C}[\Gamma_\lambda]$.

Finally, the union $\bigcup_{\lambda \in X_*(T)} \Gamma_\lambda$ is Zariski dense in $\mathfrak{t} \times \mathfrak{t} \times \mathbb{A}^1$. Hence α_{loc} is injective, and α is injective as well.

To finish the proof it suffices to compare the graded dimensions of the LHS and the RHS (the grading in the LHS arises from the natural action of $\mathbb{C}^\times$ on $\mathfrak{t}$ and $\mathbb{A}^1$). Now $\dim_{\mathrm{gr}}(H^\bullet_{G_\mathcal{O} \rtimes \mathbb{C}^\times}(\mathrm{Gr}_G)) = \dim_{\mathrm{gr}}(H^\bullet_{G_\mathcal{O}}(\mathrm{pt}) \otimes H^\bullet_{\mathbb{C}^\times}(\mathrm{pt}) \otimes H^\bullet(\mathrm{Gr}_G)) = \dim_{\mathrm{gr}} \mathbb{C}[x_1, \ldots, x_r, y_1, \ldots, y_r, \hbar]$ where $\deg \hbar = 2$, $\deg x_i = \deg y_i = 2(m_i + 1)$, and $m_1, \ldots, m_r$ are the exponents of G (so that $m_i + 1$ are the degrees of generators of W-invariant polynomials on $\mathfrak{t}$).

Furthermore, $\mathfrak{t}/W = \Sigma$, and $\mathcal{N}_\Delta(\Sigma \times \Sigma) \simeq \Sigma \times \Sigma \times \mathbb{A}^1$ (an affine space). Indeed, for affine spaces V, V' we have an isomorphism $\beta \colon V \times V' \times \mathbb{A}^1 \xrightarrow{\sim} \mathcal{N}_V(V \times V')$; namely, $\gamma \colon V \times V' \times \mathbb{A}^1 \to V \times V' \times \mathbb{A}^1$, $(v, v', c) \mapsto (v, cv', c)$ factors through $V \times V' \times \mathbb{A}^1 \xrightarrow{\beta} \mathcal{N}_{V \times V'} V \to V \times V' \times \mathbb{A}^1$. We conclude that $\dim_{\mathrm{gr}}(\mathrm{LHS}) = \dim_{\mathrm{gr}}(\mathrm{RHS})$. This completes the proof. □

In view of Proposition 1.13, we can view $H^\bullet_{G_\mathcal{O} \rtimes \mathbb{C}^\times}(\mathrm{Gr}_G, -)$ as a functor from the full subcategory $\mathcal{IC}_{G_\mathcal{O} \rtimes \mathbb{C}^\times} \subset D^b_{G_\mathcal{O} \rtimes \mathbb{C}^\times}(\mathrm{Gr}_G)$ formed by the semisimple complexes (i.e. finite direct sums of objects of the form $\mathrm{IC}(\overline{\mathrm{Gr}}^\lambda_G)[a]$), to $\mathrm{Coh}^{\mathbb{C}^\times}(\mathcal{N}_\Delta(\mathfrak{t}/W \times \mathfrak{t}/W))$. This functor is fully faithful according to [26]. For $V \in \mathrm{Rep}(G^\vee)$, one can identify $H^\bullet_{G_\mathcal{O} \rtimes \mathbb{C}^\times}(\mathrm{Gr}_G, S(V))$ with $\varkappa_\hbar(\mathrm{Fr}(V))$; moreover, one can make this identification compatible with the tensor structures on $\mathrm{Rep}(G^\vee)$ and $\mathrm{Perv}_{G_\mathcal{O}}(\mathrm{Gr}_G)$ [6]:

Theorem 1.14 *The geometric Satake equivalence* $\mathbb{S} \colon \mathrm{Rep}(G^\vee) \xrightarrow{\sim} \mathrm{Perv}_{G_\mathcal{O}}(\mathrm{Gr}_G)$ *extends to a tensor equivalence* $\mathbb{S}_\hbar \colon \mathcal{HC}^{\mathrm{fr}}_\hbar \xrightarrow{\sim} \mathcal{IC}_{G_\mathcal{O} \rtimes \mathbb{C}^\times}$ *such that* $\varkappa_\hbar = H^\bullet_{G_\mathcal{O} \rtimes \mathbb{C}^\times}(\mathrm{Gr}_G, -) \circ \mathbb{S}_\hbar$. *There is also a quasiclassical version* $\mathbb{S}_{qc} \colon \mathrm{Coh}^{G^\vee \times \mathbb{C}^\times}_{\mathrm{fr}}(\mathfrak{g}^\vee)^* \xrightarrow{\sim} \mathcal{IC}_{G_\mathcal{O}}$ *such that* $\varkappa = H^\bullet_{G_\mathcal{O}}(\mathrm{Gr}_G, -) \circ \mathbb{S}_{qc}$.

Now using the formality of RHom-algebras in our categories, one can deduce the desired derived geometric Satake equivalence. To formulate it, we introduce a bit of notation. To a dg-algebra A one can associate the triangulated category of dg-modules localized by quasi-isomorphisms, and a full triangulated subcategory $D_{\mathrm{perf}}(A) \subset D(A)$ of perfect complexes. Given an algebraic group H acting on A, we can consider H-equivariant dg-modules and localize them by quasi-isomorphisms, arriving at the equivariant version $D^H_{\mathrm{perf}}(A)$.

We now consider the dg-versions $\mathrm{Sym}^{[]}(\mathfrak{g}^\vee)$, $U_\hbar^{[]}$ of the graded algebras $\mathrm{Sym}(\mathfrak{g}^\vee)$, $U_\hbar$, equipping them with the zero differential and the cohomological grading so that elements of $\mathfrak{g}^\vee$ and $\hbar$ have degree 2. The construction of the previous paragraph gives rise to the categories $D^{G^\vee}_{\mathrm{perf}}(U_\hbar^{[]})$, $D^{G^\vee}_{\mathrm{perf}}(\mathrm{Sym}^{[]}(\mathfrak{g}^\vee))$. Their Ind-completions will be denoted by $D^{G^\vee}(U_\hbar^{[]})$, $D^{G^\vee}(\mathrm{Sym}^{[]}(\mathfrak{g}^\vee))$. The Ind-completions of the equivariant derived categories $D^b_{G_\mathcal{O}\rtimes\mathbb{C}^\times}(\mathrm{Gr}_G)$, $D^b_{G_\mathcal{O}}(\mathrm{Gr}_G)$ will be denoted by $D_{G_\mathcal{O}\rtimes\mathbb{C}^\times}(\mathrm{Gr}_G)$, $D_{G_\mathcal{O}}(\mathrm{Gr}_G)$.

The following theorem is proved in [6].

Theorem 1.15 *The equivalences of Theorem 1.14 extend to the equivalences of monoidal triangulated categories* $\Psi_\hbar\colon D^{G^\vee}_{\mathrm{perf}}(U_\hbar^{[]}) \xrightarrow{\sim} D^b_{G_\mathcal{O}\rtimes\mathbb{C}^\times}(\mathrm{Gr}_G)$ *and* $\Psi_{qc}\colon D^{G^\vee}_{\mathrm{perf}}(\mathrm{Sym}^{[]}(\mathfrak{g}^\vee)) \xrightarrow{\sim} D^b_{G_\mathcal{O}}(\mathrm{Gr}_G)$. *They induce the equivalences of their Ind-completions* $\Psi_\hbar\colon D^{G^\vee}(U_\hbar^{[]}) \xrightarrow{\sim} D_{G_\mathcal{O}\rtimes\mathbb{C}^\times}(\mathrm{Gr}_G)$ *and* $\Psi_{qc}\colon D^{G^\vee}(\mathrm{Sym}^{[]}(\mathfrak{g}^\vee)) \xrightarrow{\sim} D_{G_\mathcal{O}}(\mathrm{Gr}_G)$.

1.3.5.1 The Dualities

We denote by $\mathfrak{C}_{G^\vee}$ the autoequivalence of $D^{G^\vee}(U_\hbar^{[]})$ induced by the canonical outer automorphism of $G^\vee$ interchanging conjugacy classes of g and g^{-1} (the Chevalley involution). We also denote by $\mathfrak{C}_G$ the autoequivalence of $D_{G_\mathcal{O}\rtimes\mathbb{C}^\times}(\mathrm{Gr}_G)$ induced by $g \mapsto g^{-1}$, $G((z)) \to G((z))$. Then the Verdier duality $\mathbb{D}\colon D_{G_\mathcal{O}\rtimes\mathbb{C}^\times}(\mathrm{Gr}_G) \to D_{G_\mathcal{O}\rtimes\mathbb{C}^\times}(\mathrm{Gr}_G)$ and the duality $D\colon D^{G^\vee}(U_\hbar^{[]}) \to D^{G^\vee}(U_\hbar^{[]})$ introduced in Example 1.3.2.1 are related by $\Psi_\hbar \circ \mathfrak{C}_{G^\vee} \circ D = \mathbb{D} \circ \Psi_\hbar$.

1.3.6 The Regular Sheaf

Recall the setup of Example 1.3.2.1. We consider $\mathfrak{D}_\hbar^{[]}(G^\vee) = U_\hbar^{[]} \ltimes \mathbb{C}[G^\vee] \in D^{G^\vee}(U_\hbar^{[]})$. Its image under the equivalence of Theorem 1.15 is the *regular sheaf* $\mathcal{A}_R^{\mathbb{C}^\times} \in D_{G_\mathcal{O}\rtimes\mathbb{C}^\times}(\mathrm{Gr}_G)$ isomorphic to $\bigoplus_{\lambda\in\Lambda^+} \mathrm{IC}(\overline{\mathrm{Gr}}{}_G^\lambda) \otimes (V^\lambda)^*$. The quasiclassical analogs are $D^{G^\vee}(\mathrm{Sym}^{[]}(\mathfrak{g}^\vee)) \ni \mathbb{C}[T^*G^\vee]^{[]} = \mathrm{Sym}^{[]}(\mathfrak{g}^\vee) \otimes \mathbb{C}[G^\vee] \mapsto \mathcal{A}_R \in D_{G_\mathcal{O}}(\mathrm{Gr}_G)$. One can check that the image of $\varkappa_\hbar \mathfrak{D}_\hbar^{[]}(G^\vee) \in D^{G^\vee}(U_\hbar^{[]})$ under the equivalence of Theorem 1.15 is the dualizing sheaf $\omega_{\mathrm{Gr}_G} \in D_{G_\mathcal{O}\rtimes\mathbb{C}^\times}(\mathrm{Gr}_G)$.

It follows that the convolution algebra of equivariant Borel–Moore homology $H_\bullet^{G_\mathcal{O}\rtimes\mathbb{C}^\times}(\mathrm{Gr}_G) = H^\bullet_{G_\mathcal{O}\rtimes\mathbb{C}^\times}(\mathrm{Gr}_G, \omega_{\mathrm{Gr}_G})$ is isomorphic to the quantum open Toda lattice $\mathfrak{T}_\hbar = \varkappa_\hbar^{\mathrm{right}}\varkappa_\hbar^{\mathrm{left}}\mathfrak{D}_\hbar(G^\vee)$ of Example 1.3.2.1.

Note that the regular sheaf $\mathcal{A}_R^{\mathbb{C}^\times}$ is equipped with an action of $G^\vee$. Furthermore, the dg-algebra $\mathrm{RHom}_{D_{G_\mathcal{O}\rtimes\mathbb{C}^\times}(\mathrm{Gr}_G)}(\mathcal{A}_R^{\mathbb{C}^\times}, \mathcal{A}_R^{\mathbb{C}^\times})$ is formal, and we have a $G^\vee$-equivariant morphism of dg-algebras $U_\hbar^{[]} \to \mathrm{RHom}_{D_{G_\mathcal{O}\rtimes\mathbb{C}^\times}(\mathrm{Gr}_G)}(\mathcal{A}_R^{\mathbb{C}^\times}, \mathcal{A}_R^{\mathbb{C}^\times})$ (corresponding to the *right* action of $U_\hbar^{[]}$ on $\mathfrak{D}_\hbar^{[]}(G^\vee)$). Similarly, the dg-algebra $\mathrm{RHom}_{D_{G_\mathcal{O}}(\mathrm{Gr}_G)}(\mathcal{A}_R, \mathcal{A}_R)$ is formal, and we have a $G^\vee$-equivariant morphism of dg-algebras $\mathrm{Sym}^{[]}(\mathfrak{g}^\vee) \to \mathrm{RHom}_{D_{G_\mathcal{O}}(\mathrm{Gr}_G)}(\mathcal{A}_R, \mathcal{A}_R)$ (corresponding to the *right* action of $\mathrm{Sym}^{[]}(\mathfrak{g}^\vee)$ on $\mathbb{C}[T^*G^\vee]^{[]}$). Hence for any $\mathcal{F} \in D_{G_\mathcal{O}\rtimes\mathbb{C}^\times}(\mathrm{Gr}_G)$, the complex $\mathrm{RHom}_{D_{G_\mathcal{O}\rtimes\mathbb{C}^\times}(\mathrm{Gr}_G)}(\mathcal{A}_R^{\mathbb{C}^\times}, \mathcal{F})$ carries a structure of $G^\vee$-equivariant dg-module over $U_\hbar^{[]}$.

Thus the functors $\mathrm{RHom}_{D_{G_\mathcal{O}\rtimes\mathbb{C}^\times}(\mathrm{Gr}_G)}(\mathcal{A}_R^{\mathbb{C}^\times}, \bullet)$, $\mathrm{RHom}_{D_{G_\mathcal{O}}(\mathrm{Gr}_G)}(\mathcal{A}_R, \bullet)$ may be viewed as landing respectively into $D^{G^\vee}(U_\hbar^{[]})$, $D^{G^\vee}(\mathrm{Sym}^{[]}(\mathfrak{g}^\vee))$. We will also need their versions

$$\Phi_\hbar := \mathrm{RHom}_{D_{G_\mathcal{O}\rtimes\mathbb{C}^\times}(\mathrm{Gr}_G)}(\mathbf{1}_{\mathrm{Gr}_G}, \mathfrak{C}_G\mathcal{A}_R^{\mathbb{C}^\times} \star \bullet) \xrightarrow{\sim} \mathrm{RHom}_{D_{G_\mathcal{O}\rtimes\mathbb{C}^\times}(\mathrm{Gr}_G)}(\mathbb{D}\mathcal{A}_R^{\mathbb{C}^\times}, \bullet)$$

$$\xrightarrow{\sim} \mathrm{RHom}_{D_{G_\mathcal{O}\rtimes\mathbb{C}^\times}(\mathrm{Gr}_G)}(\mathbb{C}_{\mathrm{Gr}_G}, \mathcal{A}_R^{\mathbb{C}^\times} \overset{!}{\otimes} \bullet),$$

and

$$\Phi_{qc} := \mathrm{RHom}_{D_{G_\mathcal{O}}(\mathrm{Gr}_G)}(\mathbf{1}_{\mathrm{Gr}_G}, \mathfrak{C}_G\mathcal{A}_R \star \bullet) \xrightarrow{\sim} \mathrm{RHom}_{D_{G_\mathcal{O}}(\mathrm{Gr}_G)}(\mathbb{D}\mathcal{A}_R, \bullet)$$

$$\xrightarrow{\sim} \mathrm{RHom}_{D_{G_\mathcal{O}}(\mathrm{Gr}_G)}(\mathbb{C}_{\mathrm{Gr}_G}, \mathcal{A}_R \overset{!}{\otimes} \bullet).$$

The following lemma is proved in [15].

Lemma 1.16

(a) *The functors* $\mathrm{RHom}_{D_{G_\mathcal{O}\rtimes\mathbb{C}^\times}(\mathrm{Gr}_G)}(\mathcal{A}_R^{\mathbb{C}^\times}, \bullet)\colon D_{G_\mathcal{O}\rtimes\mathbb{C}^\times}(\mathrm{Gr}_G) \to D^{G^\vee}(U_\hbar^{[]})$ *and* $\mathrm{RHom}_{D_{G_\mathcal{O}}(\mathrm{Gr}_G)}(\mathcal{A}_R, \bullet)\colon D_{G_\mathcal{O}}(\mathrm{Gr}_G) \to D^{G^\vee}(\mathrm{Sym}^{[]}(\mathfrak{g}^\vee))$ *are canonically isomorphic to* $\Psi_\hbar^{-1}$ *and* Ψ_{qc}^{-1} *respectively.*

(b) *The functors* $\Phi_\hbar\colon D_{G_\mathcal{O}\rtimes\mathbb{C}^\times}(\mathrm{Gr}_G) \to D^{G^\vee}(U_\hbar^{[]})$ *and* $\Phi_{qc}\colon D_{G_\mathcal{O}}(\mathrm{Gr}_G) \to D^{G^\vee}(\mathrm{Sym}^{[]}(\mathfrak{g}^\vee))$ *are canonically isomorphic to* $\mathfrak{C}_{G^\vee} \circ \Psi_\hbar^{-1}$ *and* $\mathfrak{C}_{G^\vee} \circ \Psi_{qc}^{-1}$ *respectively.*

1.4 Motivation II: What do We (as Mathematicians) Might Want from 3d $N = 4$ SUSY QFT? (Naive Approach)

1.4.1 Some Generalities

In this section we would like to introduce some language related to 3-dimensional $N = 4$ super-symmetric quantum field theories. The reader should be warned from the very beginning: here we are going to use all the words from physics as a "black box". More precisely, we are not going to try to explain what such a theory really is from a mathematical point of view. Instead we are going to review the relevant "input data" (i.e. to what mathematical structures physicists usually attach such a QFT) and some of the "output data" (i.e. what mathematical structures one should get in the end). This will be largely extended in Sect. 1.7, where we partly address the question "what kind of structures these 3d $N = 4$ SUSY QFTs really are from a mathematical point of view?". Also it will be important for us to recall (in this section) what one can do with these theories: i.e. we are going to discuss some operations which produce new quantum field theories out of old ones.

The reader should be warned from the very beginning about the following: both in this section and in Sect. 1.7 we are only going to discuss algebraic aspects of the above quantum field theories (such as e.g. algebraic varieties or categories one can attach to them). In principle "true physical theory" is supposed to have some interesting analytic aspects (such as e.g. a metric on the above varieties). These analytic aspects will be completely ignored in these notes. Essentially, this means that we are going to study quantum field theories up to certain "algebraic equivalence" but we are not going to discuss details in these notes.

1.4.2 Higgs and Coulomb Branch and 3d Mirror Symmetry

A 3d $N = 4$ super-symmetric quantum field theory $\mathcal{T}$ is supposed to have a well-defined moduli space of vacua. This should be some complicated (though interesting) space. This space is somewhat too complicated for our present discussion. Instead we are going to discuss some "easy" parts of this space. Namely, the above moduli space of vacua should have two special pieces called the Higgs and the Coulomb branch of the moduli space of vacua; we shall denote these by $\mathcal{M}_H(\mathcal{T})$ and $\mathcal{M}_C(\mathcal{T})$. They are supposed to be Poisson (generically symplectic) complex algebraic varieties.[4] They should also be hyper-kähler in some sense, but (to the best of our knowledge) this notion is not well-defined for singular varieties, we are going to ignore the hyper-kähler structure in these notes. So, in this section we are

[4]In fact, this is already a simplification: non-algebraic holomorphic symplectic manifolds should also arise in this way, but we are not going to discuss such theories.

going to think about a theory $\mathcal{T}$ in terms of $\mathcal{M}_H(\mathcal{T})$ and $\mathcal{M}_C(\mathcal{T})$. Of course, this is a very small part of what the actual "physical theory" is, but we shall see that even listing the structures that physicists expect from $\mathcal{M}_H(\mathcal{T})$ and $\mathcal{M}_C(\mathcal{T})$ will lead us to interesting constructions.

One of the operations on theories that will be important in the future is the operation of "3-dimensional mirror symmetry". Namely, physicists expect that for a theory $\mathcal{T}$ there should exist a mirror dual theory $\mathcal{T}^*$ such that $\mathcal{M}_H(\mathcal{T}^*) = \mathcal{M}_C(\mathcal{T})$ and $\mathcal{M}_C(\mathcal{T}^*) = \mathcal{M}_H(\mathcal{T})$.

1.4.3 More Operations on Theories

In what follows we shall use the following notation: for a symplectic variety X with a Hamiltonian G-action we shall denote by $X /\!\!/\!\!/ G$ the Hamiltonian reduction of X with respect to G.

Then the following operations on theories are expected to make sense (in the next subsection we shall start considering examples):

1.4.3.1 Product

If $\mathcal{T}_1, \cdots, \mathcal{T}_n$ are some theories then one can form their product $\mathcal{T}_1 \times \cdots \times \mathcal{T}_n$. We have

$$\mathcal{M}_H(\mathcal{T}_1 \times \cdots \times \mathcal{T}_n) = \mathcal{M}_H(\mathcal{T}_1) \times \cdots \times \mathcal{M}_H(\mathcal{T}_n),$$

and

$$\mathcal{M}_C(\mathcal{T}_1 \times \cdots \times \mathcal{T}_n) = \mathcal{M}_C(\mathcal{T}_1) \times \cdots \times \mathcal{M}_C(\mathcal{T}_n).$$

1.4.3.2 Gauging

Let $\mathcal{T}$ be a theory and let G be a complex reductive group. Then there is a notion of G acting on $\mathcal{T}$. Physicists say in this case that G maps to the flavor symmetry group of $\mathcal{T}$, or that we are given a theory $\mathcal{T}$ with flavor symmetry G.

Assume that we are given a theory $\mathcal{T}$ with flavor symmetry G. Then there is a new theory $\mathcal{T}/G$ obtained by "gauging" G. The origin of the notation is explained in Sect. 1.4.3.3.

1.4.3.3 Result of Gauging on the Higgs Branch

Let us now address the following question: what kind of structures does a G-action on $\mathcal{T}$ imply on $\mathcal{M}_H(\mathcal{T})$ and $\mathcal{M}_C(\mathcal{T})$ and how to construct the Higgs and

Coulomb branch of $\mathcal{T}/G$? It turns out that the answer for the Higgs branch is quite straightforward but for the Coulomb branch it is much trickier. In this subsection we are going to discuss the Higgs branch and we shall postpone the discussion of the Coulomb branch till the next subsubsection.

Namely, an action of G on the theory $\mathcal{T}$ should give rise to a Hamiltonian action of G on $\mathcal{M}_H(\mathcal{T})$. Moreover, we have $\mathcal{M}_H(\mathcal{T}/G) = \mathcal{M}_H(\mathcal{T})/\!\!/\!\!/G$. Of course, the notion of Hamiltonian reduction can be understood in several different ways, so we need to talk a little about what we mean by $/\!\!/\!\!/G$ here. Recall that the Hamiltonian reduction is defined as follows. Let $\mathcal{X}$ be any Poisson variety endowed with a Hamiltonian action of G. Then we have the moment map $\mu : \mathcal{X} \to \mathfrak{g}^*$. Then we are supposed to have $\mathcal{X}/\!\!/\!\!/G = (\mu^{-1}(0))/G$. Here there are two delicate points. First, the map μ might not be flat, so honestly we must take $\mu^{-1}(0)$ in the dg-sense. Second, we must specify what we mean by quotient by G. In these notes we shall mostly deal with examples when $\mathcal{X}$ is affine and we shall be primarily interested in the algebra of functions on $\mathcal{X}/\!\!/\!\!/G$. For these purposes it is enough to work with the so called "categorical quotient", i.e. we set

$$\mathbb{C}[\mathcal{X}/\!\!/\!\!/G] = (\mathbb{C}[\mu^{-1}(0)])^G.$$

Note that according to our conventions this might be a dg-algebra.

1.4.3.4 Ring Object

Given $\mathcal{T}$ and G as above (assuming that G is connected and reductive) one can construct a ring object $\mathcal{A}_\mathcal{T}$ in $D_{G_\mathcal{O}}(\mathrm{Gr}_G)$ (sometimes we shall denote it by $\mathcal{A}_{\mathcal{T},G}$ when we need to stress the dependence on G). The !-stalk of $\mathcal{A}_\mathcal{T}$ at the unit point of Gr_G is $\mathbb{C}[\mathcal{M}_C(\mathcal{T})]$ and $H^*_{G_\mathcal{O}}(\mathcal{A}_\mathcal{T}) = \mathbb{C}[\mathcal{M}_C(\mathcal{T}/G)]$. In fact, the object $\mathcal{A}_\mathcal{T}$ should also have a Poisson structure (which will induce a Poisson structure on $\mathbb{C}[\mathcal{M}_C(\mathcal{T})]$ and on $\mathcal{M}_C(\mathcal{T}/G)$) but we are going to ignore this issue for now.

Let us as before denote by i the emebedding of the point 1 in Gr_G. Then $i^!\mathcal{A}_\mathcal{T}$ can be regarded as a ring object of the equivariant derived category $D_G(\mathrm{pt})$. Its equivariant cohomology $H^*_G(\mathrm{pt}, i^!\mathcal{A}_\mathcal{T})$ is a graded algebra over $H^*_G(\mathrm{pt}, \mathbb{C}) = \mathbb{C}[\mathfrak{g}]^G = \mathbb{C}[\mathfrak{t}]^W$ whose (derived) fiber over 0 is equal to $\mathbb{C}[\mathcal{M}_C]$. Thus flavor symmetry G is supposed to define a (Poisson) deformation of $\mathcal{M}_C$ over the base $\mathfrak{t}/W$. In particular, by base change we should have a Poisson deformation of $\mathcal{M}_C$ over $\mathfrak{t}$.

1.4.3.5 Ring Object for a Subgroup

Let H be a subgroup of G. Then $\mathcal{A}_{\mathcal{T},H}$ is equal to the !-restriction of $\mathcal{A}_{\mathcal{T},G}$ to Gr_H.

1.4.3.6 A Theory $\mathcal{T}[G]$

For a reductive group G there is a theory $\mathcal{T}[G]$ such that

(a) $\mathcal{T}[G]$ has flavor symmetry G.
(b) $\mathcal{M}_H(\mathcal{T}[G]) = \mathcal{N}_{\mathfrak{g}}$, $\mathcal{M}_C(\mathcal{T}[G]) = \mathcal{N}_{\mathfrak{g}^\vee}$; here $\mathfrak{g} = \mathrm{Lie}(G)$ and $\mathcal{N}_{\mathfrak{g}}$ is its nilpotent cone.
(c) $\mathcal{A}_{\mathcal{T}[G^\vee]} = \mathcal{A}_R$ (our "regular representation" sheaf on Gr_G).
(d) $\mathcal{T}[G]^* = \mathcal{T}[G^\vee]$.

1.4.3.7 S-Duality

For a theory $\mathcal{T}$ with flavor symmetry G there should exist another (S-dual) theory $\mathcal{T}^\vee$ with flavor symmetry $G^\vee$ (it is defined via S-duality for 4-dimensional $N = 4$ super-symmetric gauge theory). Gaiotto and Witten [20] claim that

$$\mathcal{T}^\vee = ((\mathcal{T} \times \mathcal{T}[G])/G)^* \tag{1.2}$$

(here the gauging is taken with respect to diagonal copy of G).

In particular, the RHS of (1.2) has an action of $G^\vee$ (which a priori is absolutely non-obvious). Here is an example: take $\mathcal{T}$ to be the trivial theory with trivial G-action (in this case $\mathcal{M}_H = \mathcal{M}_C = \mathrm{pt}$, but the structure is still somewhat non-trivial as we remember the group G). Then $\mathcal{T} \times \mathcal{T}[G] = \mathcal{T}[G]$. Now $\mathcal{T}[G]/G$ is the theory whose (naive, i.e. not dg) Higgs branch is pt and whose Coulomb branch is isomorphic to $G^\vee \times \mathfrak{t}/W = \mathrm{Spec}(H^*_{G_\mathcal{O}}(\mathrm{Gr}_G, \mathcal{A}_R))$. Since the mirror duality interchanges $\mathcal{M}_H$ and $\mathcal{M}_C$ we see that $\mathcal{M}_H(\mathcal{T}^\vee)$ has an action of $G^\vee$.

More generally, it follows that

$$\mathbb{C}[\mathcal{M}_H(\mathcal{T}^\vee)] = H^*_{G_\mathcal{O}}(\mathrm{Gr}_G, \mathcal{A}_\mathcal{T} \otimes \mathcal{A}_R) \tag{1.3}$$

(this follows from Sect. 1.4.3.5). In particular, it has a natural action of $G^\vee$.

Let us now pass to examples.

1.4.4 Basic Example

This is in some sense the most basic example. Let $\mathbf{M}$ be a connected symplectic algebraic variety over $\mathbb{C}$. Then to $\mathbf{M}$ there should correspond a theory $T(\mathbf{M})$ for which $\mathcal{M}_H = \mathbf{M}$ and $\mathcal{M}_C = \mathrm{pt}$. In fact, this is true only if dg-structures are disregarded. However, in these notes we shall mostly care about the case when $\mathbf{M}$ is just a symplectic vector space and in this case it should be true as stated (cf. Sect. 1.7.11 for more details).

1.4.5 Gauge Theories

Let G be a reductive group. Then an action of G on $T(\mathbf{M})$ should be the same as a Hamiltonian action of G on $\mathbf{M}$.[5] Then we can form the theory $T(\mathbf{M})/G$.

Assume that $\mathbf{M}$ is actually a symplectic vector space and that the action of G on $\mathbf{M}$ is linear. Then the theory $T(\mathbf{M})/G$ is called *gauge theory with matter* $\mathbf{M}$. In the case when $\mathbf{M} = \mathbf{N} \oplus \mathbf{N}^* = T^*\mathbf{N}$ for some repesentation $\mathbf{N}$ of G, the Coulomb branch of these theories together with the corresponding objects $\mathcal{A}_{\mathcal{T}}$ was rigorously defined and studied in [13–15]. Unfortunately, at this point we don't know how to modify our constructions so that they will depend on $\mathbf{M}$ rather than on $\mathbf{N}$ (but we can check that different ways of representing $\mathbf{M}$ as $T^*\mathbf{N}$ (in the cases where it is possible) lead to the same $\mathcal{M}_C$). We shall sometimes denote the theory $T(\mathbf{M})/G$ simply by $T(G, \mathbf{N})$.

Here is an interesting source of pairs $(G, \mathbf{N})$ as above. Let Q be an oriented quiver (a.k.a. finite oriented graph) with set of vertices I. Let V and W be two finite-dimensional I-graded vector spaces over $\mathbb{C}$. Set

$$G = \prod_{i \in I} \mathrm{GL}(V_i), \quad \mathbf{N} = (\bigoplus_{i \to j} \mathrm{Hom}(V_i, V_j)) \oplus (\bigoplus \mathrm{Hom}(V_i, W_i)).$$

Theories associated with such pairs $(G, \mathbf{N})$ are called *quiver gauge theories*. In the case when Q is a quiver of finite Dynkin type the corresponding Coulomb branches are studied in detail in [14]; we review some of these results in Sect. 1.6.

1.4.6 Toric Gauge Theories

Let $T \subset (\mathbb{C}^\times)^n$ be an algebraic torus. We set $T_F = (\mathbb{C}^\times)^n/T$ (this is also an algebraic torus). Then T acts naturally on $\mathbb{C}^n$, so we can set $\mathbf{N} = \mathbb{C}^n$, $G = T$ in the notation of the previous subsection.

Note that the torus $T_F^\vee$ also naturally embeds to $(\mathbb{C}^\times)^n$ (by dualizing the quotient map $(\mathbb{C}^\times)^n \to T_F$). It is then expected that the mirror dual to the theory associated to $(T, \mathbb{C}^n)$ is equal to the theory associated with $(T_F^\vee, \mathbb{C}^n)$. Note that this implies that the Coulomb branch of the former must be isomorphic to $T^*\mathbb{C}^n /\!\!/\!\!/ T_F^\vee$ (since this is the Higgs branch of the latter). As was mentioned earlier, in the next section we are going to give a rigorous definition of Coulomb branches for gauge theories of cotangent type and the above expectation in the toric case is proven in Example 1.5.4.1.

---

[5]By Hamiltonian action we mean a symplectic action with fixed moment map.

1.4.7 Sicilian Theories

Let Σ be a Riemann surface obtained from a compact Riemann surface $\overline{\Sigma}$ by making n punctures. Let also G be a reductive group. To this data physicists associate a theory $T(\Sigma, G)$ ("Sicilian theory") with an action of G^n. The construction is by "compactifying" certain 6-dimensional theory (attached to G) on $\Sigma \times S^1$.

One of the key statements from physics is that the theory associated to a sphere with n-punctures and the group $G^\vee$ is

$$((\underbrace{\mathcal{T}[G] \times \cdots \times \mathcal{T}[G]}_{n \text{ times}}) / G)^*. \tag{1.4}$$

Here we are gauging the diagonal action of G. It has an action of $(G^\vee)^n$ for reasons similar to Sect. 1.4.3.7. There should be in fact a simpler statement (when you start with a theory corresponding to any surface and make an additional puncture), but we are a little confused now about what it is. In particular, it says that functions on the Higgs branch (for any G) of (1.4) is $H^*_{G_{\mathcal{O}}}(\mathrm{Gr}_G, \underbrace{\mathcal{A}_R \overset{!}{\otimes} \cdots \overset{!}{\otimes} \mathcal{A}_R}_{n \text{ times}})$ (which is what we knew before for G of type A). More precisely, for $G = GL(r)$ the theory $\mathcal{T}[G]$ is a quiver theory of type A_{r+1} and then the theory $\underbrace{(\mathcal{T}[G] \times \cdots \times \mathcal{T}[G])}_{n \text{ times}}/G$ is the corresponding star-shaped quiver theory. Interested reader can consult [15, Section 6] for more details.

1.4.8 S-Duality vs. Derived Satake

Let $\mathcal{T}$ be a theory with G-symmetry and let $\mathcal{A}_{\mathcal{T},G}$ be the corresponding ring object on Gr_G. We would like to describe the corresponding data for the S-dual theory $\mathcal{T}^\vee$. Let Ψ_{qc}^{-1} denote the derived geometric Satake functor going from $D(\mathrm{Gr}_G)$ to the derived category of $G^\vee$-equivariant dg-modules over $\mathrm{Sym}(\mathfrak{g}^\vee[-2])$ (see Theorem 1.14). Then the cohomology $h^*(\Psi_{qc}^{-1}(\mathcal{A}_{\mathcal{T},G}))$ with grading disregarded can be viewed as a commutative ring object in the cateogory of $G^\vee$-equivariant modules over $\mathrm{Sym}(\mathfrak{g}^\vee)$. In other words, $\mathrm{Spec}(h^*(\Psi_{qc}^{-1}(\mathcal{A}_{\mathcal{T},G})))$ is a $G^\vee$-scheme endowed with a compatible morphism to $(\mathfrak{g}^\vee)^*$.

It follows from the results of the previous section that

$$\mathcal{M}_H(\mathcal{T}^\vee) = \mathrm{Spec}(h^*(\Psi_{qc}^{-1}(\mathcal{A}_{\mathcal{T},G}))).$$

Another (categorical) relationship between the assignment $\mathcal{T} \mapsto \mathcal{T}^\vee$ and geometric Langlands duality will be discussed in Sect. 1.7.16.

1.5 Coulomb Branches of 3-Dimensional Gauge Theories

In this section we explain how to define Coulomb branches (and some further structures related to flavor symmetry) for gauge theories of cotangent type.

1.5.1 Summary

Let us summarize what is done in this section. Let G be a complex connected reductive group and let $\mathbf{N}$ be a representation of it. In this section we are going to define mathematically the Coulomb branch $\mathcal{M}_C(G, \mathbf{N})$ of the gauge theory $\mathcal{T}(T^*\mathbf{N})/G$. These Coulomb branches will satisfy the following (non-exhaustive) list of properties:

(1) $\mathcal{M}_C(G, \mathbf{N})$ is a normal, affine generically symplectic Poisson variety (conjecturally it is singular symplectic but we don't know how to prove this).
(2) Let T be a maximal torus in G and let W be the Weyl group of G. Then $\mathcal{M}_C(G, \mathbf{N})$ is birationally isomorphic to $(T^*T^\vee)/W$. This birational isomorphism is compatible with the Poisson structure. In particular, $\dim(\mathcal{M}_C(G, \mathbf{N}) = 2\operatorname{rank} G$.
(3) There is a natural "integrable system" map $\pi : \mathcal{M}_C(G, \mathbf{N}) \to \mathfrak{t}/W$ which has Lagrangian fibers.
(4) $\mathcal{M}_C(G, \mathbf{N})$ is equipped with a canonical quantization; the map π also gets quantized.

1.5.2 General Setup

Let $\mathbf{N}$ be a finite dimensional representation of a complex connected reductive group G. We consider the moduli space $\mathcal{R}_{G,\mathbf{N}}$ of triples $(\mathcal{P}, \sigma, s)$ where $\mathcal{P}$ is a G-bundle on the formal disc $D = \operatorname{Spec} \mathcal{O}$; σ is a trivialization of $\mathcal{P}$ on the punctured formal disc $D^* = \operatorname{Spec} \mathcal{K}$; and s is a section of the associated vector bundle $\mathcal{P}_{\text{triv}} \overset{G}{\times} \mathbf{N}$ on D^* such that s extends to a regular section of $\mathcal{P}_{\text{triv}} \overset{G}{\times} \mathbf{N}$ on D, and $\sigma(s)$ extends to a regular section of $\mathcal{P} \overset{G}{\times} \mathbf{N}$ on D. In other words, s extends to a regular section of the vector bundle associated to the G-bundle glued from $\mathcal{P}$ and $\mathcal{P}_{\text{triv}}$ on the non-separated formal scheme glued from 2 copies of D along D^* (*raviolo*). The group $G_\mathcal{O}$ acts on $\mathcal{R}_{G,\mathbf{N}}$ by changing the trivialization σ, and we have an evident $G_\mathcal{O}$-equivariant projection $\mathcal{R}_{G,\mathbf{N}} \to \mathrm{Gr}_G$ forgetting s. The fibers of this projection are profinite dimensional vector spaces: the fiber over the base point is $\mathbf{N} \otimes \mathcal{O}$, and all the other fibers are subspaces in $\mathbf{N} \otimes \mathcal{O}$ of finite codimension. One may say that $\mathcal{R}_{G,\mathbf{N}}$ is a $G_\mathcal{O}$-equvariant "constructible profinite dimensional vector bundle" over Gr_G.

1.5.2.1 Example: Affine Steinberg Variety

If $\mathbf{N}$ is the adjoint representation $G \curvearrowright \mathfrak{g}$, then $\mathcal{R}_{G,\mathbf{N}}$ is isomorphic to the union $\bigcup_{\lambda \in \Lambda^+} T^*_{\mathrm{Gr}^\lambda_G} \mathrm{Gr}_G$ of conormal bundles to the $G_\mathcal{O}$-orbits in Gr_G.

The $G_\mathcal{O}$-equivariant Borel–Moore homology $H^{G_\mathcal{O}}_\bullet(\mathcal{R}_{G,\mathbf{N}})$ is defined via the following limiting procedure.

We define $\mathcal{R}_{\leq\lambda}$ as the preimage of $\overline{\mathrm{Gr}}^\lambda_G$ in $\mathcal{R} := \mathcal{R}(G, \mathbf{N})$. It suffices to define the $G_\mathcal{O}$-equivariant Borel–Moore homology $H^{G_\mathcal{O}}_\bullet(\mathcal{R}_{\leq\lambda})$ along with the maps $H^{G_\mathcal{O}}_\bullet(\mathcal{R}_{\leq\lambda}) \to H^{G_\mathcal{O}}_\bullet(\mathcal{R}_{\leq\mu})$ for $\lambda \leq \mu$. For a fixed λ and $d \gg 0$, $\mathcal{R}_{\leq\lambda}$ is invariant under the translations by $z^d\mathbf{N}_\mathcal{O}$, and we denote the quotient by $\mathcal{R}^d_{\leq\lambda}$, so that $\mathcal{R}_{\leq\lambda} = \varprojlim \mathcal{R}^d_{\leq\lambda}$. For fixed λ, d, and $e \gg 0$, the action of $G_\mathcal{O}$ on $\mathcal{R}^d_{\leq\lambda}$ factors through the action of $G_{\mathcal{O}/z^e\mathcal{O}}$. Finally,

$$H^{G_\mathcal{O}}_\bullet(\mathcal{R}_{\leq\lambda}) := H^{-\bullet}_{G_{\mathcal{O}/z^e\mathcal{O}}}(\mathcal{R}^d_{\leq\lambda}, \omega_{\mathcal{R}^d_{\leq\lambda}})[-2\dim(\mathbf{N}_\mathcal{O}/z^d\mathbf{N}_\mathcal{O})].$$

The cohomological shift means that we are considering the "renormalized" Borel–Moore homology, i.e. the cohomology $H^{-\bullet}_{G_\mathcal{O}}(\mathcal{R}, \omega_\mathcal{R}[-2\dim \mathbf{N}_\mathcal{O}])$.

The $G_\mathcal{O}$-equivariant Borel–Moore homology $H^{G_\mathcal{O}}_\bullet(\mathcal{R}_{G,\mathbf{N}})$ forms an associative algebra with respect to the following convolution operation. We consider the diagram

$$\begin{array}{ccccccc}
\mathcal{R}\times\mathcal{R} & \xleftarrow{\tilde{p}} & p^{-1}(\mathcal{R}\times\mathcal{R}) & \xrightarrow{\tilde{q}} & q(p^{-1}(\mathcal{R}\times\mathcal{R})) & \xrightarrow{\tilde{m}} & \mathcal{R} \\
{\scriptstyle i\times\mathrm{Id}_\mathcal{R}}\downarrow & & {\scriptstyle i'}\downarrow & & \downarrow & & \downarrow{\scriptstyle i} \\
\mathcal{T}\times\mathcal{R} & \xleftarrow{p} & G_\mathcal{K}\times\mathcal{R} & \xrightarrow{q} & G_\mathcal{K}\overset{G_\mathcal{O}}{\times}\mathcal{R} & \xrightarrow{m} & \mathcal{T},
\end{array} \tag{1.5}$$

Here $\mathcal{T} := G_\mathcal{K} \overset{G_\mathcal{O}}{\times} \mathbf{N}_\mathcal{O}$, and we have an embedding $\mathcal{T} \hookrightarrow \mathrm{Gr}_G \times \mathbf{N}_\mathcal{K}$ such that $\mathcal{R} = \mathcal{T} \cap (\mathrm{Gr}_G \times \mathbf{N}_\mathcal{O})$. The embedding $\mathcal{R} \hookrightarrow \mathcal{T}$ is denoted by i. The maps in the lower row are given by

$$\big(g_1, [g_2, s]\big) \overset{q}{\mapsto} \big[g_1, [g_2, s]\big] \overset{m}{\mapsto} [g_1g_2, s], \ \big(g_1, [g_2, s]\big) \overset{p}{\mapsto} \big([g_1, g_2 s], [g_2, s]\big),$$

and all the squares are cartesian (i.e. the upper row consists of closed subvarietes in the lower row, and all the maps in the upper row are induced by the corresponding maps in the lower row). We have the following group actions on the terms of the lower row preserving the closed subvarieties in the upper row:

$$G_\mathcal{O} \times G_\mathcal{O} \curvearrowright \mathcal{T} \times \mathcal{R};\ (g, h) \cdot ([g_1, s_1], [g_2, s_2]) = ([gg_1, s_1], [hg_2, s_2]),$$

$$G_\mathcal{O} \times G_\mathcal{O} \curvearrowright G_\mathcal{K} \times \mathcal{R};\ (g, h) \cdot (g_1, [g_2, s]) = \Big(gg_1h^{-1}, [hg_2, s]\Big),$$

$$G_{\mathcal{O}} \curvearrowright G_{\mathcal{K}} \overset{G_{\mathcal{O}}}{\times} \mathcal{R};\ g \cdot [g_1, [g_2, s]] = [gg_1, [g_2, s]],$$

$$G_{\mathcal{O}} \curvearrowright \mathcal{T};\ g \cdot [g_1, s] = [gg_1, s].$$

The morphisms p, q, m (and hence $\tilde{p}, \tilde{q}, \tilde{m}$) are equivariant, where we take the first projection $\mathrm{pr}_1 : G_{\mathcal{O}} \times G_{\mathcal{O}} \to G_{\mathcal{O}}$ for q.

Finally, given two equivariant Borel–Moore homology classes $c_1, c_2 \in H_\bullet^{G_{\mathcal{O}}}(\mathcal{R})$, we define their convolution product $c_1 * c_2 := \tilde{m}_*(\tilde{q}^*)^{-1}\tilde{p}^*(c_1 \otimes c_2)$.

This algebra is commutative, finitely generated and integral, and its spectrum $\mathcal{M}_C(G, \mathbf{N}) = \mathrm{Spec}\, H_\bullet^{G_{\mathcal{O}}}(\mathcal{R}_{G,\mathbf{N}})$ is an irreducible normal affine variety of dimension $2\,\mathrm{rk}(G)$, the *Coulomb branch*. It is supposed to be a (singular) hyper-Kähler manifold [46].

Let $T \subset G$ be a Cartan torus with Lie algebra $\mathfrak{t} \subset \mathfrak{g}$. Let $W = N_G(T)/T$ be the corresponding Weyl group. Then the equivariant cohomology $H^\bullet_{G_{\mathcal{O}}}(\mathrm{pt}) = \mathbb{C}[\mathfrak{t}/W]$ forms a subalgebra of $H_\bullet^{G_{\mathcal{O}}}(\mathcal{R}_{G,\mathbf{N}})$ (a *Cartan subalgebra*), so we have a projection $\Pi\colon \mathcal{M}_C(G, \mathbf{N}) \to \mathfrak{t}/W$.

1.5.2.2 Example

For the adjoint representation $\mathbf{N} = \mathfrak{g}$ considered in Sect. 1.5.2.1, we get $\mathcal{M}_C(G, \mathfrak{g}) = (T^\vee \times \mathfrak{t})/W$. For the trivial representation, we get $\mathcal{M}_C(G, 0) = \mathfrak{Z}^{G^\vee}_{\mathfrak{g}^\vee} = \{(g \in G^\vee, \xi \in \Sigma) : \mathrm{Ad}_g \xi = \xi\}$, the universal centralizer of the dual group. Compare with Proposition 1.13 where the spectrum of the equivariant *cohomology* of the affine Grassmannian is computed.

Finally, the algebra $H_\bullet^{G_{\mathcal{O}}}(\mathcal{R}_{G,\mathbf{N}})$ comes equipped with quantization: a $\mathbb{C}[\hbar]$-deformation $\mathbb{C}_\hbar[\mathcal{M}_C(G, \mathbf{N})] = H_\bullet^{G_{\mathcal{O}} \rtimes \mathbb{C}^\times}(\mathcal{R}_{G,\mathbf{N}})$ where $\mathbb{C}^\times$ acts by loop rotations, and $\mathbb{C}[\hbar] = H^\bullet_{\mathbb{C}^\times}(\mathrm{pt})$. It gives rise to a Poisson bracket on $\mathbb{C}[\mathcal{M}_C(G, \mathbf{N})]$ with an open symplectic leaf, so that Π becomes an integrable system: $\mathbb{C}[\mathfrak{t}/W] \subset \mathbb{C}[\mathcal{M}_C(G, \mathbf{N})]$ is a Poisson-commutative polynomial subalgebra with $\mathrm{rk}(G)$ generators.

1.5.3 Monopole Formula

Recall that $\mathcal{R}_{G,\mathbf{N}}$ is a union of (profinite dimensional) vector bundles over $G_{\mathcal{O}}$-orbits in Gr_G. The corresponding Cousin spectral sequence converging to $H_\bullet^{G_{\mathcal{O}}}(\mathcal{R}_{G,\mathbf{N}})$ degenerates and allows to compute the equivariant Poincaré polynomial (or rather Hilbert series)

$$P_t^{G_{\mathcal{O}}}(\mathcal{R}_{G,\mathbf{N}}) = \sum_{\theta \in \Lambda^+} t^{d_\theta - 2\langle \rho^\vee, \theta \rangle} P_G(t; \theta). \tag{1.6}$$

Here $\deg(t) = 2$, $P_G(t;\theta) = \prod(1-t^{d_i})^{-1}$ is the Hilbert series of the equivariant cohomology $H^\bullet_{\mathrm{Stab}_G(\theta)}(\mathrm{pt})$ (d_i are the degrees of generators of the ring of $\mathrm{Stab}_G(\theta)$-invariant functions on its Lie algebra), and $d_\theta = \sum_{\chi\in\Lambda_G^\vee} \max(-\langle\chi,\theta\rangle, 0) \dim \mathbf{N}_\chi$. This is a slight variation of the *monopole formula* of [17]. Note that the series (1.6) may well diverge (even as a formal Laurent series: the space of homology of given degree may be infinite-dimensional), e.g. this is always the case for unframed quiver gauge theories. To ensure its convergence (as a formal Taylor series with the constant term 1) one has to impose the so called 'good' or 'ugly' assumption on the theory. In this case the resulting $\mathbb{N}$-grading on $H^{G_\mathcal{O}}_\bullet(\mathcal{R}_{G,\mathbf{N}})$ gives rise to a $\mathbb{C}^\times$-action on $\mathcal{M}_C(G,\mathbf{N})$, making it a conical variety with a single (attracting) fixed point.

1.5.4 Flavor Symmetry

Suppose we have an extension $1 \to G \to \tilde{G} \to G_F \to 1$ where G_F is a connected reductive group (a *flavor group*), and the action of G on $\mathbf{N}$ is extended to an action of $\tilde{G}$. Then the action of $G_\mathcal{O}$ on $\mathcal{R}_{G,\mathbf{N}}$ extends to an action of $\tilde{G}_\mathcal{O}$, and the convolution product defines a commutative algebra structure on the equivariant Borel–Moore homology $H^{\tilde{G}_\mathcal{O}}_\bullet(\mathcal{R}_{G,\mathbf{N}})$. We have the restriction homomorphism $H^{\tilde{G}_\mathcal{O}}_\bullet(\mathcal{R}_{G,\mathbf{N}}) \to H^{G_\mathcal{O}}_\bullet(\mathcal{R}_{G,\mathbf{N}}) = H^{\tilde{G}_\mathcal{O}}_\bullet(\mathcal{R}_{G,\mathbf{N}}) \otimes_{H^\bullet_{G_F}(\mathrm{pt})} \mathbb{C}$. In other words, $\underline{\mathcal{M}}_C(G,\mathbf{N}) := \mathrm{Spec}\, H^{\tilde{G}_\mathcal{O}}_\bullet(\mathcal{R}_{G,\mathbf{N}})$ is a deformation of $\mathcal{M}_C(G,\mathbf{N})$ over $\mathrm{Spec}\, H^\bullet_{G_F}(\mathrm{pt}) = \mathfrak{t}_F/W_F$.

We will need the following version of this construction. Let $\mathsf{Z} \subset G_F$ be a torus embedded into the flavor group. We denote by $\tilde{G}^\mathsf{Z}$ the pullback extension $1 \to G \to \tilde{G}^\mathsf{Z} \to \mathsf{Z} \to 1$. We define $\underline{\mathcal{M}}^\mathsf{Z}_C(G,\mathbf{N}) := \mathrm{Spec}\, H^{\tilde{G}^\mathsf{Z}_\mathcal{O}}_\bullet(\mathcal{R}_{G,\mathbf{N}})$: a deformation of $\mathcal{M}_C(G,\mathbf{N})$ over $\mathfrak{z} := \mathrm{Spec}\, H^\bullet_\mathsf{Z}(\mathrm{pt})$.

Since $\mathcal{M}_C(G,\mathbf{N})$ is supposed to be a hyper-Kähler manifold, its flavor deformation should come together with a (partial) resolution. To construct it, we consider the obvious projection $\tilde{\pi}\colon \mathcal{R}_{\tilde{G},\mathbf{N}} \to \mathrm{Gr}_{\tilde{G}} \to \mathrm{Gr}_{G_F}$. Given a dominant coweight $\lambda_F \in \Lambda_F^+ \subset \mathrm{Gr}_{G_F}$, we set $\mathcal{R}^{\lambda_F}_{\tilde{G},\mathbf{N}} := \tilde{\pi}^{-1}(\lambda_F)$, and consider the equivariant Borel–Moore homology $H^{\tilde{G}^\mathsf{Z}_\mathcal{O}}_\bullet(\mathcal{R}^{\lambda_F}_{\tilde{G},\mathbf{N}})$. It carries a convolution module structure over $H^{\tilde{G}^\mathsf{Z}_\mathcal{O}}_\bullet(\mathcal{R}_{G,\mathbf{N}})$. We consider $\underline{\widetilde{\mathcal{M}}}^{\mathsf{Z},\lambda_F}_C(G,\mathbf{N}) := \mathrm{Proj}(\bigoplus_{n\in\mathbb{N}} H^{\tilde{G}^\mathsf{Z}_\mathcal{O}}_\bullet(\mathcal{R}^{n\lambda_F}_{\tilde{G},\mathbf{N}})) \xrightarrow{\varpi} \underline{\mathcal{M}}^\mathsf{Z}_C(G,\mathbf{N})$. We denote $\varpi^{-1}(\mathcal{M}_C(G,\mathbf{N}))$ by $\widetilde{\mathcal{M}}^{\lambda_F}_C(G,\mathbf{N})$. We have $\widetilde{\mathcal{M}}^{\lambda_F}_C(G,\mathbf{N}) = \mathrm{Proj}(\bigoplus_{n\in\mathbb{N}} H^{G_\mathcal{O}}_\bullet(\mathcal{R}^{n\lambda_F}_{\tilde{G},\mathbf{N}}))$.

More generally, for a strictly convex (i.e. not containing nontrivial subgroups) cone $\mathsf{V} \subset \Lambda_F^+$, we consider the multi projective spectra $\underline{\widetilde{\mathcal{M}}}^{\mathsf{Z},\mathsf{V}}_C(G,\mathbf{N}) := \mathrm{Proj}(\bigoplus_{\lambda_F\in\mathsf{V}} H^{\tilde{G}^\mathsf{Z}_\mathcal{O}}_\bullet(\mathcal{R}^{\lambda_F}_{\tilde{G},\mathbf{N}})) \xrightarrow{\varpi} \underline{\mathcal{M}}^\mathsf{Z}_C(G,\mathbf{N})$ and $\widetilde{\mathcal{M}}^\mathsf{V}_C(G,\mathbf{N}) := \mathrm{Proj}(\bigoplus_{\lambda_F\in\mathsf{V}} H^{G_\mathcal{O}}_\bullet(\mathcal{R}^{\lambda_F}_{\tilde{G},\mathbf{N}})) \xrightarrow{\varpi} \mathcal{M}_C(G,\mathbf{N})$.

The following proposition is proved in [13].

Proposition 1.17 *Assume that the flavor group is a torus, i.e. we have an exact sequence* $1 \to G \to \tilde{G} \to T_F \to 1$. *Then the Coulomb branch* $\mathcal{M}_C(G, \mathbf{N})$ *is the Hamiltonian reduction of* $\mathcal{M}_C(\tilde{G}, \mathbf{N})$ *by the action of the dual torus* $T_F^\vee$.

1.5.4.1 Example: Toric Hyper-Kähler Manifolds

Consider an exact sequence

$$0 \to \mathbb{Z}^{d-n} \xrightarrow{\alpha} \mathbb{Z}^d \xrightarrow{\beta} \mathbb{Z}^n \to 0$$

and the associated sequence

$$1 \to G = (\mathbb{C}^\times)^{d-n} \xrightarrow{\alpha} \tilde{G} = (\mathbb{C}^\times)^d \xrightarrow{\beta} T_F = (\mathbb{C}^\times)^n \to 1 \tag{1.7}$$

Let $\mathbf{N} = \mathbb{C}^d$ considered as a representation of G via α. By Proposition 1.17, the Coulomb branch $\mathcal{M}_C(G, \mathbf{N})$ is the Hamiltonian reduction of $\mathcal{M}_C((\mathbb{C}^\times)^d, \mathbb{C}^d)$ by the action of $T_F^\vee$. It is easy to see that $\mathcal{M}_C((\mathbb{C}^\times)^d, \mathbb{C}^d) = \mathcal{M}_C(\mathbb{C}^\times, \mathbb{C})^d \simeq \mathbb{A}^{2d}$, and hence $\mathcal{M}_C(G, \mathbf{N})$ is, by definition, the toric hyper-Kähler manifold associated with the dual sequence of (1.7) [7].

In particular, if $\mathbf{N}$ is a 1-dimensional representation of $\mathbb{C}^\times$ with the character q^n, then $\mathcal{M}_C(\mathbb{C}^\times, \mathbf{N})$ is the Kleinian surface of type A_{n-1} given by the equation $xy = w^n$. If $\mathbf{N}$ is an n-dimensional representation of $\mathbb{C}^\times$ with the character nq, then the Coulomb branch $\mathcal{M}_C(\mathbb{C}^\times, \mathbf{N})$ is the same Kleinian surface of type A_{n-1}.

1.5.5 Ring Objects in the Derived Satake Category

Let π stand for the projection $\mathcal{R} \to \mathrm{Gr}_G$. Then $\mathcal{A}^{\mathbb{C}^\times} := \pi_*\omega_{\mathcal{R}}[-2\dim \mathbf{N}_\mathcal{O}]$ is an object of $D_{G_\mathcal{O} \rtimes \mathbb{C}^\times}(\mathrm{Gr}_G)$, and $H^\bullet_{G_\mathcal{O} \rtimes \mathbb{C}^\times}(\mathcal{R}, \omega_{\mathcal{R}}[-2\dim \mathbf{N}_\mathcal{O}]) = H^\bullet_{G_\mathcal{O} \rtimes \mathbb{C}^\times}(\mathrm{Gr}_G, \mathcal{A})$. One can equip $\mathcal{A}^{\mathbb{C}^\times}$ with a structure of a ring object in $D_{G_\mathcal{O} \rtimes \mathbb{C}^\times}(\mathrm{Gr}_G)$ so that the resulting ring structure on $H^\bullet_{G_\mathcal{O} \rtimes \mathbb{C}^\times}(\mathrm{Gr}_G, \mathcal{A}^{\mathbb{C}^\times})$ coincides with the ring structure on $H_\bullet^{G_\mathcal{O} \rtimes \mathbb{C}^\times}(\mathcal{R})$ introduced in Sect. 1.5.2. If we forget the loop rotation equivariance, then the resulting ring object $\mathcal{A}$ of $D_{G_\mathcal{O}}(\mathrm{Gr}_G)$ is commutative.

Similarly, in the situation of Sect. 1.5.4, we denote $\tilde{\mathcal{R}} := \mathcal{R}(\tilde{G}, \mathbf{N})$, and consider the composed projection $\tilde{\pi} : \tilde{\mathcal{R}} \to \mathrm{Gr}_{\tilde{G}} \to \mathrm{Gr}_{G_F}$. We define a ring object $\mathcal{A}_F^{\mathbb{C}^\times} := \mathrm{Ind}_{\tilde{G}_\mathcal{O} \rtimes \mathbb{C}^\times}^{(G_F)_\mathcal{O} \rtimes \mathbb{C}^\times} \tilde{\pi}_*\omega_{\tilde{\mathcal{R}}}[-2\dim \mathbf{N}_\mathcal{O}] \in D_{(G_F)_\mathcal{O} \rtimes \mathbb{C}^\times}(\mathrm{Gr}_{G_F})$, where $\mathrm{Ind}_{\tilde{G}_\mathcal{O} \rtimes \mathbb{C}^\times}^{(G_F)_\mathcal{O} \rtimes \mathbb{C}^\times}$ is the functor changing equivariance from $\tilde{G}_\mathcal{O} \rtimes \mathbb{C}^\times$ to $(G_F)_\mathcal{O} \rtimes \mathbb{C}^\times$. If we forget the loop rotation equivariance, we obtain a commutative ring object $\mathcal{A}_F \in D_{(G_F)_\mathcal{O}}(\mathrm{Gr}_{G_F})$. We will also need the fully equivariant ring object $\tilde{\mathcal{A}}_F^{\mathbb{C}^\times} := \tilde{\pi}_*\omega_{\tilde{\mathcal{R}}}[-2\dim \mathbf{N}_\mathcal{O}] \in D_{\tilde{G}_\mathcal{O} \rtimes \mathbb{C}^\times}(\mathrm{Gr}_{G_F})$.

The ring $\mathbb{C}_\hbar[\mathcal{M}_C(G,\mathbf{N})]$ is reconstructed from the ring object $\tilde{\mathcal{A}}_F^{\mathbb{C}^\times}$ by the following procedure going back to [3]. For a flavor coweight λ_F we denote by i_{λ_F} the embedding of a T_F-fixed point λ_F into Gr_{G_F}. Then $\mathrm{Ext}^\bullet_{D_{\tilde{G}_\mathcal{O}\rtimes\mathbb{C}^\times}(\mathrm{Gr}_{G_F})}(\mathbf{1}_{\mathrm{Gr}_{G_F}}, \tilde{\mathcal{A}}_F^{\mathbb{C}^\times}) = i_0^!\tilde{\mathcal{A}}_F^{\mathbb{C}^\times} \simeq H^\bullet_{\tilde{G}_\mathcal{O}\rtimes\mathbb{C}^\times}(\mathcal{R}, \omega_\mathcal{R}[-2\dim\mathbf{N}_\mathcal{O}])$ by the base change. Given $x, y \in \mathrm{Ext}^\bullet_{D_{\tilde{G}_\mathcal{O}\rtimes\mathbb{C}^\times}(\mathrm{Gr}_{G_F})}(\mathbf{1}_{\mathrm{Gr}_{G_F}}, \tilde{\mathcal{A}}_F^{\mathbb{C}^\times})$, we consider $x \star y \in \mathrm{Ext}^\bullet_{D_{\tilde{G}_\mathcal{O}\rtimes\mathbb{C}^\times}(\mathrm{Gr}_{G_F})}(\mathbf{1}_{\mathrm{Gr}_{G_F}} \star \mathbf{1}_{\mathrm{Gr}_{G_F}}, \tilde{\mathcal{A}}_F^{\mathbb{C}^\times} \star \tilde{\mathcal{A}}_F^{\mathbb{C}^\times})$, and then apply the isomorphism $\mathbf{1}_{\mathrm{Gr}_{G_F}} \simeq \mathbf{1}_{\mathrm{Gr}_{G_F}} \star \mathbf{1}_{\mathrm{Gr}_{G_F}}$ and the multiplication morphism $\mathsf{m}\colon \tilde{\mathcal{A}}_F^{\mathbb{C}^\times} \star \tilde{\mathcal{A}}_F^{\mathbb{C}^\times} \to \tilde{\mathcal{A}}_F^{\mathbb{C}^\times}$ in order to obtain $\mathsf{m}(x \star y) \in \mathrm{Ext}^\bullet_{D_{\tilde{G}_\mathcal{O}\rtimes\mathbb{C}^\times}(\mathrm{Gr}_{G_F})}(\mathbf{1}_{\mathrm{Gr}_{G_F}}, \tilde{\mathcal{A}}_F^{\mathbb{C}^\times})$. It is proved in [15] that the resulting ring structure on $\mathrm{Ext}^\bullet_{D_{\tilde{G}_\mathcal{O}\rtimes\mathbb{C}^\times}(\mathrm{Gr}_{G_F})}(\mathbf{1}_{\mathrm{Gr}_{G_F}}, \tilde{\mathcal{A}}_F^{\mathbb{C}^\times}) = H_\bullet^{\tilde{G}_\mathcal{O}\rtimes\mathbb{C}^\times}(\mathcal{R})$ induces the one introduced in Sect. 1.5.2 on $H_\bullet^{G_\mathcal{O}\rtimes\mathbb{C}^\times}(\mathcal{R})$. Moreover, a similar construction defines a multiplication $i_{\lambda_F}^!\bar{\mathcal{A}}_F^{\mathbb{C}^\times} \otimes i_{\mu_F}^!\bar{\mathcal{A}}_F^{\mathbb{C}^\times} \to i_{\lambda_F+\mu_F}^!\bar{\mathcal{A}}_F^{\mathbb{C}^\times}$ for $\lambda_F, \mu_F \in \Lambda_F^+$. Here $\bar{\mathcal{A}}_F^{\mathbb{C}^\times} = \mathrm{Res}^{G_\mathcal{O}\rtimes\mathbb{C}^\times}_{\tilde{G}_\mathcal{O}\rtimes\mathbb{C}^\times}\tilde{\mathcal{A}}_F^{\mathbb{C}^\times}$ is obtained from $\tilde{\mathcal{A}}_F^{\mathbb{C}^\times}$ applying the functor restricting equivariance from $\tilde{G}_\mathcal{O} \rtimes \mathbb{C}^\times$ to $G_\mathcal{O} \rtimes \mathbb{C}^\times$. In particular, we get a module structure $i_0^!\bar{\mathcal{A}}_F^{\mathbb{C}^\times} \otimes i_{\lambda_F}^!\bar{\mathcal{A}}_F^{\mathbb{C}^\times} \to i_{\lambda_F}^!\bar{\mathcal{A}}_F^{\mathbb{C}^\times}$. Note that $i_0^!\bar{\mathcal{A}}_F^{\mathbb{C}^\times} \simeq H_\bullet^{G_\mathcal{O}\rtimes\mathbb{C}^\times}(\mathcal{R})$.

1.5.5.1 Example: The Regular Sheaf in Type A

Let $G = \mathrm{GL}(\mathbb{C}^{N-1}) \times \mathrm{GL}(\mathbb{C}^{N-2}) \times \ldots \times \mathrm{GL}(\mathbb{C}^1)$, $\tilde{G} = (G \times \mathrm{GL}(\mathbb{C}^N))/Z$, where $Z \simeq \mathbb{C}^\times$ is the diagonal central subgroup. Hence $G_F = \mathrm{PGL}(\mathbb{C}^N)$. Furthermore, $\mathbf{N} = \mathrm{Hom}(\mathbb{C}^N, \mathbb{C}^{N-1}) \oplus \mathrm{Hom}(\mathbb{C}^{N-1}, \mathbb{C}^{N-2}) \oplus \ldots \oplus \mathrm{Hom}(\mathbb{C}^2, \mathbb{C}^1)$. It is proved in [15] that $\mathcal{A}_F^{\mathbb{C}^\times}$ is isomorphic to the regular sheaf $\mathcal{A}_R^{\mathbb{C}^\times} \in D_{\mathrm{PGL}(\mathbb{C}^N)_\mathcal{O}\rtimes\mathbb{C}^\times}(\mathrm{Gr}_{\mathrm{PGL}(\mathbb{C}^N)})$ of Sect. 1.3.6.

1.5.6 Gluing Construction

Let $\mathcal{A}_1^{\mathbb{C}^\times}, \ldots, \mathcal{A}_n^{\mathbb{C}^\times}$ be the ring objects in $D_{G_\mathcal{O}\rtimes\mathbb{C}^\times}(\mathrm{Gr}_G)$. We denote the ring objects of $D_{G_\mathcal{O}}(\mathrm{Gr}_G)$ obtained by forgetting the loop rotation equivariance by $\mathcal{A}_1, \ldots, \mathcal{A}_n$. Let $i_\Delta\colon \mathrm{Gr}_G \hookrightarrow \prod_{k=1}^n \mathrm{Gr}_G$ be the diagonal embedding. The following proposition is proved in [15].

Proposition 1.18 *$\mathcal{A}^{\mathbb{C}^\times} := i_\Delta^!(\boxtimes\mathcal{A}_k^{\mathbb{C}^\times})$ is a ring object in $D_{G_\mathcal{O}\rtimes\mathbb{C}^\times}(\mathrm{Gr}_G)$. If the ring objects $\mathcal{A}_1, \ldots, \mathcal{A}_n$ are commutative, then $\mathcal{A} := i_\Delta^!(\boxtimes\mathcal{A}_k) \in D_{G_\mathcal{O}}(\mathrm{Gr}_G)$ is a commutative ring object. In particular, the ring $H^\bullet_{G_\mathcal{O}}(\mathrm{Gr}_G, \mathcal{A})$ is commutative.*

Proof We have $\boxtimes\mathsf{m}\colon (\boxtimes\mathcal{A}_k) \star (\boxtimes\mathcal{A}_k) = \boxtimes(\mathcal{A}_k \star \mathcal{A}_k) \to \boxtimes\mathcal{A}_k$ from $\mathsf{m}\colon \mathcal{A}_k \star \mathcal{A}_k \to \mathcal{A}_k$. Then we apply $i_\Delta^!$. We claim that there is a natural homomorphism

$$i_\Delta^!(\boxtimes\mathcal{A}_k) \star i_\Delta^!(\boxtimes\mathcal{A}_k) \to i_\Delta^!\left(\boxtimes(\mathcal{A}_k \star \mathcal{A}_k)\right),$$

hence its composition with $i^!_\Delta(\boxtimes\mathfrak{m})$ gives the desired multiplication homomorphism of $i^!_\Delta(\boxtimes\mathcal{A}_k)$. We prove the claim by comparing the convolution diagrams (1.1) for Gr_G and $\prod_k \mathrm{Gr}_G$. Since p, q are smooth, p^*, q^* commute with $i^!_\Delta$. The last part of the convolution diagram for G and $\prod_k G$ is

$$\begin{array}{ccc} \mathrm{Gr}_G \tilde{\times} \mathrm{Gr}_G & \xrightarrow{m} & \mathrm{Gr}_G \\ {\scriptstyle i'_\Delta}\big\downarrow & & \big\downarrow{\scriptstyle i_\Delta} \\ \prod_k \mathrm{Gr}_G \tilde{\times} \mathrm{Gr}_G = \mathrm{Gr}_{\prod_k G} \tilde{\times} \mathrm{Gr}_{\prod_k G} & \xrightarrow[\prod_k m]{} & \mathrm{Gr}_{\prod_k G} = \prod_k \mathrm{Gr}_G, \end{array}$$

where we denote the diagonal embedding of the left column by i'_Δ to distinguish it from the right column. Let $\boxtimes(\mathcal{A}_k \tilde{\boxtimes} \mathcal{A}_k)$ denote the complex on $\mathrm{Gr}_{\prod_k G} \tilde{\times} \mathrm{Gr}_{\prod_k G}$ obtained in the course of the convolution product for $\prod_k G$. We define the homomorphism as

$$m_* i'^!_\Delta(\boxtimes(\mathcal{A}_k \tilde{\boxtimes} \mathcal{A}_k)) = m_* \overset{!}{\bigotimes} (\mathcal{A}_k \tilde{\boxtimes} \mathcal{A}_k) \to$$

$$\overset{!}{\bigotimes} m_*(\mathcal{A}_k \tilde{\boxtimes} \mathcal{A}_k) = i^!_\Delta(\prod_k m)_* \boxtimes (\mathcal{A}_k \tilde{\boxtimes} \mathcal{A}_k)).$$

□

Recall the regular sheaf $\mathcal{A}_R^{\mathbb{C}^\times}$ of Sect. 1.3.6. It is equipped with an action of $G^\vee \ltimes U_\hbar^{[]}$. Hence for any ring object $\mathcal{A}^{\mathbb{C}^\times} \in D_{G_\mathcal{O} \rtimes \mathbb{C}^\times}(\mathrm{Gr}_G)$, the product $\mathcal{A}_R^{\mathbb{C}^\times} \otimes^! \mathcal{A}^{\mathbb{C}^\times}$ is also equipped with an action of $G^\vee \ltimes U_\hbar^{[]}$. The cohomology ring $H^\bullet_{G_\mathcal{O} \rtimes \mathbb{C}^\times}(\mathrm{Gr}_G, \mathcal{A}_R^{\mathbb{C}^\times} \otimes^! \mathcal{A}^{\mathbb{C}^\times})$ is also equipped with an action of $G^\vee \ltimes U_\hbar^{[]}$. The following proposition is proved in [15] (recall that the autoequivalence $\mathfrak{C}_{G^\vee}$ was defined in Sect. 1.3.5.1):

Proposition 1.19 *For ring objects $\mathcal{A}_1^{\mathbb{C}^\times}, \mathcal{A}_2^{\mathbb{C}^\times} \in D_{G_\mathcal{O} \rtimes \mathbb{C}^\times}(\mathrm{Gr}_G)$, we have*

$$\begin{aligned} &H^\bullet_{G_\mathcal{O} \rtimes \mathbb{C}^\times}(\mathrm{Gr}_G, \mathcal{A}_1^{\mathbb{C}^\times} \overset{!}{\otimes} \mathcal{A}_2^{\mathbb{C}^\times}) \\ &\simeq H^\bullet_{G_\mathcal{O} \rtimes \mathbb{C}^\times}(\mathrm{Gr}_G, \mathcal{A}_R^{\mathbb{C}^\times} \overset{!}{\otimes} \mathcal{A}_1^{\mathbb{C}^\times}) \otimes \mathfrak{C}_{G^\vee} H^\bullet_{G_\mathcal{O} \rtimes \mathbb{C}^\times}(\mathrm{Gr}_G, \mathcal{A}_R^{\mathbb{C}^\times} \overset{!}{\otimes} \mathcal{A}_2^{\mathbb{C}^\times}) /\!/\!/ \Delta_{G^\vee} \end{aligned}$$

(quantum Hamiltonian reduction). If the ring objects $\mathcal{A}_1, \mathcal{A}_2 \in D_{G_\mathcal{O}}(\mathrm{Gr}_G)$ obtained by forgetting the loop rotation equivariance are commutative, then we have a similar isomorphism of commutative rings:

$$H^\bullet_{G_\mathcal{O}}(\mathrm{Gr}_G, \mathcal{A}_1 \overset{!}{\otimes} \mathcal{A}_2) \simeq H^\bullet_{G_\mathcal{O}}(\mathrm{Gr}_G, \mathcal{A}_R \overset{!}{\otimes} \mathcal{A}_1) \otimes \mathfrak{C}_{G^\vee} H^\bullet_{G_\mathcal{O}}(\mathrm{Gr}_G, \mathcal{A}_R \overset{!}{\otimes} \mathcal{A}_2) /\!/\!/ \Delta_{G^\vee}$$

Proof By rigidity, we have

$$\begin{aligned}
H^{\bullet}_{G_{\mathcal{O}}\rtimes\mathbb{C}^\times}(\mathrm{Gr}_G, \mathcal{A}_1^{\mathbb{C}^\times} \overset{!}{\otimes} \mathcal{A}_2^{\mathbb{C}^\times}) &= \mathrm{Ext}^{\bullet}_{D_{G_{\mathcal{O}}\rtimes\mathbb{C}^\times}(\mathrm{Gr}_G)}(\mathbb{D}\mathcal{A}_1^{\mathbb{C}^\times}, \mathcal{A}_2^{\mathbb{C}^\times}) \\
&= \mathrm{Ext}^{\bullet}_{D_{G_{\mathcal{O}}\rtimes\mathbb{C}^\times}(\mathrm{Gr}_G)}(\mathbf{1}_{\mathrm{Gr}_G}, \mathfrak{C}_G\mathcal{A}_1^{\mathbb{C}^\times} \star \mathcal{A}_2^{\mathbb{C}^\times}) \\
&= \mathrm{Ext}^{\bullet}_{D^{G^\vee}(U_\hbar^{[]})}\left(U_\hbar^{[]}, \mathfrak{C}_{G^\vee}\Psi_\hbar^{-1}(\mathcal{A}_1^{\mathbb{C}^\times}) \otimes_{U_\hbar^{[]}} \Psi_\hbar^{-1}(\mathcal{A}_2^{\mathbb{C}^\times})\right) \\
&= \mathrm{Ext}^{\bullet}_{D^{G^\vee}(U_\hbar^{[]})}\left(U_\hbar^{[]}, \Phi_\hbar(\mathcal{A}_1^{\mathbb{C}^\times}) \otimes_{U_\hbar^{[]}} \mathfrak{C}_{G^\vee}\Phi_\hbar(\mathcal{A}_2^{\mathbb{C}^\times})\right),
\end{aligned}$$

(the last equality is Lemma 1.16(b)). Now it is easy to see that $\mathrm{Ext}^{\bullet}_{D^{G^\vee}(U_\hbar^{[]})}\left(U_\hbar^{[]}, \Phi_\hbar(\mathcal{A}_1^{\mathbb{C}^\times}) \otimes_{U_\hbar^{[]}} \mathfrak{C}_{G^\vee}\Phi_\hbar(\mathcal{A}_2^{\mathbb{C}^\times})\right)$ is the hamiltonian reduction $(\Phi_\hbar(\mathcal{A}_1^{\mathbb{C}^\times}) \otimes \mathfrak{C}_{G^\vee}\Phi_\hbar(\mathcal{A}_2^{\mathbb{C}^\times}))/\!\!/\!\!/\Delta_{G^\vee}$ of $\Phi_\hbar(\mathcal{A}_1^{\mathbb{C}^\times}) \otimes \mathfrak{C}_{G^\vee}\Phi_\hbar(\mathcal{A}_2^{\mathbb{C}^\times})$ with respect to the diagonal action of $G^\vee$. Finally, according to Lemma 1.16, $H^{\bullet}_{G_{\mathcal{O}}\rtimes\mathbb{C}^\times}(\mathrm{Gr}_G, \mathcal{A}_R^{\mathbb{C}^\times} \overset{!}{\otimes} \mathcal{A}_{1,2}^{\mathbb{C}^\times}) = \Phi_\hbar(\mathcal{A}_{1,2}^{\mathbb{C}^\times})$. □

1.5.7 Higgs Branches of Sicilian Theories

We denote $i_\Delta^!(\mathcal{A}_R^{\boxtimes b})$ by $\mathcal{A}^b \in D_{G_{\mathcal{O}}}(\mathrm{Gr}_G)$. It is equipped with an action of b copies of $G^\vee \ltimes U_\hbar^{[]}$. We denote by $\mathcal{B} \in D_{G_{\mathcal{O}}}(\mathrm{Gr}_G)$ the quantum hamiltonian reduction of $\mathcal{A}^2$ by the diagonal action $G^\vee$. We expect that $\mathcal{B}$ is isomorphic to $\pi_*\omega_{\mathcal{R}_{G,\mathfrak{g}}}[-2\dim \mathfrak{g}_{\mathcal{O}}]$ (see Sect. 1.5.5 and Example 1.5.2.1). Finally, we set $\mathcal{B}^g := i_\Delta^!(\mathcal{B}^{\boxtimes g})$. Then $\mathcal{A}^b \otimes^! \mathcal{B}^g$ is a commutative ring object of $D_{G_{\mathcal{O}}}(\mathrm{Gr}_G)$, and its equivariant cohomology is a commutative ring. We denote by $W_G^{g,b}$ its spectrum $\mathrm{Spec}\, H^{\bullet}_{G_{\mathcal{O}}}(\mathrm{Gr}_G, \mathcal{A}^b \otimes^! \mathcal{B}^g)$. It is a Poisson variety equipped with an action of $(G^\vee)^b$, the conjectural Higgs branch of a Sicilian theory.

Recall that according to [38], there is a conjectural functor from the category of 2-bordisms to a category HS of holomorphic symplectic varieties with Hamiltonian group actions. The objects of HS are complex algebraic semisimple groups. A morphism from G to G' is a holomorphic symplectic variety X with a $\mathbb{C}^\times$-action scaling the symplectic form with weight 2, together with hamiltonian $G \times G'$-action commuting with the $\mathbb{C}^\times$-action. For $X \in \mathrm{Mor}(G', G)$, $Y \in \mathrm{Mor}(G, G'')$, the composition $Y \circ X \in \mathrm{Mor}(G', G'')$ is given by the symplectic reduction of $Y \times X$ by the diagonal G-action. The identity morphism in $\mathrm{Mor}(G, G)$ is the cotangent bundle T^*G with the left and right action of G.

To a complex semisimple group G and a Riemann surface with boundary, physicists associate a $3d$ Sicilian theory and consider its Higgs branch. It depends only on the topology of the Riemann surface, and gives a functor as above. Such a functor satisfying most of expected properties was constructed recently in [27]. It follows from Proposition 1.19 that the above $W_G^{g,b}$ is associated to the group $G^\vee$ and

Riemann surface of genus g with b boundary components. It is also proved in [15] that $W^{0,3}_{\mathrm{PGL}(2)} \simeq \mathbb{C}^2 \otimes \mathbb{C}^2 \otimes \mathbb{C}^2$, and $W^{0,3}_{\mathrm{PGL}(3)}$ is the minimal nilpotent orbit of E_6, while $W^{1,1}_{\mathrm{PGL}(3)}$ is the subregular nilpotent orbit of G_2, as expected by physicists.

1.6 Coulomb Branches of 3*d* Quiver Gauge Theories

1.6.1 Quiver Gauge Theories

Let Q be a quiver with Q_0 the set of vertices, and Q_1 the set of arrows. An arrow $e \in Q_1$ goes from its tail $t(e) \in Q_0$ to its head $h(e) \in Q_0$. We choose a Q_0-graded vector spaces $V := \bigoplus_{j\in Q_0} V_j$ and $W := \bigoplus_{j\in Q_0} W_j$. We set $\mathsf{G} = \mathrm{GL}(V) := \prod_{j\in Q_0} \mathrm{GL}(V_j)$. We choose a second grading $W = \bigoplus_{s=1}^N W^{(s)}$ compatible with the Q_0-grading of W. We set G_F to be a Levi subgroup $\prod_{s=1}^N \prod_{j\in Q_0} \mathrm{GL}(W_j^{(s)})$ of $\mathrm{GL}(W)$, and $\tilde{\mathsf{G}} := \mathsf{G} \times \mathsf{G}_F$.

Remark G will be the gauge group in this section. We denote it by G since we want to use the notation G for some other group.

Finally, we define a central subgroup $\mathsf{Z} \subset \mathsf{G}_F$ as follows: $\mathsf{Z} := \prod_{s=1}^N \Delta^{(s)}_{\mathbb{C}^\times} \subset \prod_{s=1}^N \prod_{j\in Q_0} \mathrm{GL}(W_j^{(s)})$, where $\mathbb{C}^\times \cong \Delta^{(s)}_{\mathbb{C}^\times} \subset \prod_{j\in Q_0} \mathrm{GL}(W_j^{(s)})$ is the diagonally embedded subgroup of scalar matrices. The reductive group $\tilde{\mathsf{G}}$ acts naturally on $\mathbf{N} := \bigoplus_{e\in Q_1} \mathrm{Hom}(V_{t(e)}, V_{h(e)}) \oplus \bigoplus_{j\in Q_0} \mathrm{Hom}(W_j, V_j)$.

The Higgs branch of the corresponding quiver gauge theory is the Nakajima quiver variety $\mathcal{M}_H(\mathsf{G}, \mathbf{N}) = \mathfrak{M}(V, W)$. We are interested in the Coulomb branch $\mathcal{M}_C(\mathsf{G}, \mathbf{N})$.

1.6.2 Generalized Slices in an Affine Grassmannian

Recall the slices $\overline{\mathcal{W}}^\lambda_\mu$ in the affine Grassmannian Gr_G of a reductive group G, defined in Sect. 1.2.1 for domimant μ. For arbitrary μ we consider the moduli space $\overline{\mathcal{W}}^\lambda_\mu$ of the following data:

(a) A G-bundle $\mathcal{P}$ on $\mathbb{P}^1$.
(b) A trivialization $\sigma : \mathcal{P}_{\mathrm{triv}}|_{\mathbb{P}^1\setminus\{0\}} \xrightarrow{\sim} \mathcal{P}|_{\mathbb{P}^1\setminus\{0\}}$ having a pole of degree $\leq \lambda$ at $0 \in \mathbb{P}^1$ (that is defining a point of $\overline{\mathrm{Gr}}^\lambda_G$).
(c) A B-structure ϕ on $\mathcal{P}$ of degree $w_0\mu$ with the fiber $B_- \subset G$ at $\infty \in \mathbb{P}^1$ (with respect to the trivialization σ of $\mathcal{P}$ at $\infty \in \mathbb{P}^1$). Here $G \supset B_- \supset T$ is the Borel subgroup opposite to B, and $w_0 \in W$ is the longest element.

This construction goes back to [19]. The space $\overline{\mathcal{W}}^\lambda_\mu$ is nonempty iff $\mu \leq \lambda$. In this case it is an irreducible affine normal Cohen–Macaulay variety of dimension $\langle 2\rho^\vee, \lambda - \mu \rangle$, see [14]. In case μ is dominant, the two definitions of $\overline{\mathcal{W}}^\lambda_\mu$ agree. At the other extreme, if $\lambda = 0$, then $\overline{\mathcal{W}}^0_{-\alpha}$ is nothing but the open zastava space $\mathring{Z}^{-w_0\alpha}$. The T-fixed point set $(\overline{\mathcal{W}}^\lambda_\mu)^T$ is nonempty iff the weight space V^λ_μ is not 0; in this case $(\overline{\mathcal{W}}^\lambda_\mu)^T$ consists of a single point denoted μ.

1.6.3 Beilinson-Drinfeld Slices

Let $\underline{\lambda} = (\lambda_1, \ldots, \lambda_N)$ be a collection of dominant coweights of G. We consider the moduli space $\underline{\overline{\mathcal{W}}}^{\underline{\lambda}}_\mu$ of the following data:

(a) A collection of points $(z_1, \ldots, z_N) \in \mathbb{A}^N$ on the affine line $\mathbb{A}^1 \subset \mathbb{P}^1$.
(b) A G-bundle $\mathcal{P}$ on $\mathbb{P}^1$.
(c) A trivialization $\sigma : \mathcal{P}_{\mathrm{triv}}|_{\mathbb{P}^1 \setminus \{z_1,\ldots,z_N\}} \xrightarrow{\sim} \mathcal{P}|_{\mathbb{P}^1 \setminus \{z_1,\ldots,z_N\}}$ with a pole of degree $\leq \sum_{s=1}^N \lambda_s \cdot z_s$ on the complement.
(d) A B-structure ϕ on $\mathcal{P}$ of degree $w_0\mu$ with the fiber $B_- \subset G$ at $\infty \in \mathbb{P}^1$ (with respect to the trivialization σ of $\mathcal{P}$ at $\infty \in \mathbb{P}^1$).

$\underline{\overline{\mathcal{W}}}^{\underline{\lambda}}_\mu$ is nonempty iff $\mu \leq \lambda := \sum_{s=1}^N \lambda_s$. In this case it is an irreducible affine normal Cohen–Macaulay variety flat over $\mathbb{A}^N$ of relative dimension $\langle 2\rho^\vee, \lambda - \mu \rangle$, see [14]. The fiber over $N \cdot 0 \in \mathbb{A}^N$ is nothing but $\overline{\mathcal{W}}^\lambda_\mu$.

1.6.4 Convolution Diagram Over Slices

In the setup of Sect. 1.6.3 we consider the moduli space $\underline{\widetilde{\mathcal{W}}}^{\underline{\lambda}}_\mu$ of the following data:

(a) A collection of points $(z_1, \ldots, z_N) \in \mathbb{A}^N$ on the affine line $\mathbb{A}^1 \subset \mathbb{P}^1$.
(b) A collection of G-bundles $(\mathcal{P}_1, \ldots, \mathcal{P}_N)$ on $\mathbb{P}^1$.
(c) A collection of isomorphisms $\sigma_s : \mathcal{P}_{s-1}|_{\mathbb{P}^1 \setminus \{z_s\}} \xrightarrow{\sim} \mathcal{P}_s|_{\mathbb{P}^1 \setminus \{z_s\}}$ with a pole of degree $\leq \lambda_s$ at z_s. Here $1 \leq s \leq N$, and $\mathcal{P}_0 := \mathcal{P}_{\mathrm{triv}}$.
(d) A B-structure ϕ on $\mathcal{P}_N$ of degree $w_0\mu$ with the fiber $B_- \subset G$ at $\infty \in \mathbb{P}^1$ (with respect to the trivialization $\sigma_N \circ \ldots \circ \sigma_1$ of $\mathcal{P}_N$ at $\infty \in \mathbb{P}^1$).

A natural projection $\varpi : \underline{\widetilde{\mathcal{W}}}^{\underline{\lambda}}_\mu \to \underline{\overline{\mathcal{W}}}^{\underline{\lambda}}_\mu$ sends $(\mathcal{P}_1, \ldots, \mathcal{P}_N, \sigma_1, \ldots, \sigma_N)$ to $(\mathcal{P}_N, \sigma_N \circ \ldots \circ \sigma_1)$. We denote $\varpi^{-1}(\overline{\mathcal{W}}^\lambda_\mu)$ by $\widetilde{\mathcal{W}}^{\underline{\lambda}}_\mu$. Then we expect that $\varpi : \widetilde{\mathcal{W}}^{\underline{\lambda}}_\mu \to \overline{\mathcal{W}}^\lambda_\mu$ is stratified semismall.

1.6.5 Slices as Coulomb Branches

Let now G be an adjoint simple simply laced algebraic group. We choose an orientation Ω of its Dynkin graph (of type ADE), and denote by I its set of vertices. Given an I-graded vector space W we encode its dimension by a dominant coweight $\lambda := \sum_{i\in I} \dim(W_i)\omega_i \in \Lambda^+$ of G. Given an I-graded vector space V we encode its dimension by a positive coroot combination $\alpha :\doteq \sum_{i\in I} \dim(V_i)\alpha_i \in \Lambda_+$. We set $\mu := \lambda - \alpha \in \Lambda$. Given a direct sum decomposition $W = \bigoplus_{s=1}^N W^{(s)}$ compatible with the I-grading of W as in Sect. 1.6.1, we set $\lambda_s := \sum_{i\in I} \dim(W_i^{(s)})\omega_i \in \Lambda^+$, and finally, $\underline{\lambda} := (\lambda_1, \ldots, \lambda_N)$.

Recall the notations of Sect. 1.5.4. Since the flavor group G_F is a Levi subgroup of $\mathrm{GL}(W)$, its weight lattice is naturally identified with $\mathbb{Z}^{\dim W}$. More precisely, we choose a basis $w_1, \ldots, w_{\dim W}$ of W such that any W_i, $i \in I$, and $W^{(s)}$, $1 \le s \le N$, is spanned by a subset of the basis, and we assume the following monotonicity condition: if for $1 \le a < b < c \le \dim W$ we have $w_a, w_b \in W^{(s)}$ for certain s, then $w_b \in W^{(s)}$ as well. We define a strictly convex cone $\mathsf{V} = \{(n_1, \ldots, n_{\dim W})\} \subset \Lambda_F^+ \subset \mathbb{Z}^{\dim W}$ by the following conditions: (a) if $w_k \in W^{(s)}$, $w_l \in W^{(t)}$, and $s < t$, then $n_k \ge n_l \ge 0$; (b) if $w_k, w_l \in W^{(s)}$, then $n_k = n_l$. The following theorem is proved in [14, 16] by the fixed point localization and reduction to calculations in rank 1:

Theorem 1.20 *We have isomorphisms*

$$\overline{\mathcal{W}}^{\lambda}_{\mu} \xrightarrow{\sim} \mathcal{M}_C(\mathsf{G}, \mathbf{N}),\ \underline{\overline{\mathcal{W}}}^{\underline{\lambda}}_{\mu} \xrightarrow{\sim} \underline{\mathcal{M}}^{\mathsf{Z}}_C(\mathsf{G}, \mathbf{N}),\ \underline{\widetilde{\mathcal{W}}}^{\underline{\lambda}}_{\mu} \xrightarrow{\sim} \underline{\widetilde{\mathcal{M}}}^{\mathsf{Z},\mathsf{V}}_C(\mathsf{G}, \mathbf{N}),\ \widetilde{\mathcal{W}}^{\underline{\lambda}}_{\mu} \xrightarrow{\sim} \widetilde{\mathcal{M}}^{\mathsf{V}}_C(\mathsf{G}, \mathbf{N}).$$

1.6.6 Further Examples

Let now Q be an *affine* quiver of type $\tilde{A}\tilde{D}\tilde{E}$; the framing W is 1-dimensional concentrated at the extending vertex; and the dimension of V is d times the minimal imaginary coroot δ. Then it is expected that $\mathcal{M}_C(\mathsf{G}, \mathbf{N})$ is isomorphic to the Uhlenbeck (partial) compactification $\mathcal{U}^d_G(\mathbb{A}^2)$ [12] of the moduli space of G-bundles on $\mathbb{P}^2$ trivialized at $\mathbb{P}^1_\infty$, of second Chern class d. This is proved for $G = \mathrm{SL}(N)$ in [43].

Furthermore, let Q be a star-shaped quiver with b legs of length N each, and with g loop-edges at the central vertex. The framing is trivial, and the dimension of V along each leg, starting at the outer end, is $1, 2, \ldots, N-1, N$ (with N at the central vertex). Contrary to the general setup in Sect. 1.6.1, we define G as the quotient of $\mathrm{GL}(V)$ by the diagonal central subgroup $\mathbb{C}^\times$ (acting trivially on $\mathbf{N}$). Then according to [15], $\mathcal{M}_C(\mathsf{G}, \mathbf{N})$ is isomorphic to $W^{g,b}_{\mathrm{PGL}(N)}$ of Sect. 1.5.7.

1.7 More Physics: Topological Twists of 3d $N = 4$ QFT and Categorical Constructions

The constructions of this section are mostly conjectural. The main idea of this section is given by Eq. (1.10) which is due to T. Dimofte, D. Gaiotto, J. Hilburn and P. Yoo. We discuss some interesting corollaries of this equation.

1.7.1 Extended Topological Field Theories

Physical quantum field theories usually depend on a choice of metric on the space-time. The theory is called topological if all the quantities (e.g. corelation functions) are independent of the metric (however, look at the warning at the end of the next subsection). Mathematically, the axioms of a topological QFT were first formulated by Atiyah (cf. [4]). Roughly speaking, a topological quantum field theory in dimension d consists of the following data:

(a) A complex number $Z(M^d)$ for every compact oriented d-dimensional manifold M^d;
(b) A vector space $Z(M^{d-1})$ for every compact oriented $(d-1)$-dimensional manifold M^{d-1};
(c) A vector in $Z(\partial M)$ for every compact oriented d-dimensional manifold M with boundary ∂M.

These data must satisfy certain list of standard axioms; we refer the reader to [4] for details. In addition, one can consider a richer structure called *extended topological field theory*. This structure in addition to (a), (b) and (c) as above must associate k-category $Z(M^{d-k-1})$ to a compact oriented manifold M^{d-k-1} of dimension $d-k-1$. It should also associate an object of the k-category $Z(\partial M)$ to every compact oriented manifold M of dimension $d-k$; more generally, there is a structure associated with every *manifold with corners* of dimension $\leq d$. We refer the reader to [34] for details about extended topological field theories. In the sequel we shall be mostly concerned with the case $d = 3$. In this case one is supposed to associate a (usual) category to the circle S^1. Physicists call it *the category of line operators*.

1.7.2 Topological Twists of 3d $N = 4$ Theories

Physical quantum field theories are usually not topological. However, sometimes physicists can produce a universal procedure which associates a topological field theory to a physical theory with enough super-symmetry. Since in these notes we are not discussing what a quantum field theory really is, we can't discuss what a topological twist really is. Physicists say that any 3d $N = 4$ theory with some mild additional structure[6] must have two topological twists (we'll call them Coulomb and Higgs twists, although physicists often call them A and B twists by analogy with similar construction for 2-dimensional field theories). These twists must be interchanged by the 3d mirror symmetry operation mentioned in Sect. 1.4.

1.7.3 Warning

The twists are topological only in some weak sense. Namely, in principle as was mentioned above in a topological field theory everything (e.g. correlators) should be independent of the metric (i.e. only depend on the topology of the relevant space-time). In a weakly topological field theory everything should be metric-independent only locally. This issue will be ignored in this section since we are only going to discuss some pretty robust things but it is actually important if one wants to understand some finer aspects.

1.7.4 The Category of Line Operators in a Topologically Twisted 3d $N = 4$ Theory

To a 3d TFT one should be able to attach a "category of line operators" (i.e. this is the category one attaches to a circle in terms of the previous subsection). Morever, since the circle S^1 is the boundary of a canonical 2-dimensional manifold: the 2-dimensional disc, this category should come equipped with a canonical object. In this section we would like to suggest a construction of these categories together with the above object for a wide class of topologically twisted 3d $N = 4$ theories (we learned the idea of this construction from T. Dimofte, D. Gaiotto, J. Hilburn and P. Yoo who can actually derive this construction from physical considerations. To the best of our knowledge their paper on the subject is forthcoming).

A priori the above categories of line operators should be $\mathbb{Z}_2$-graded. However, as was mentioned above, in order to define the relevant topological twists one needs to choose some mild additional structure on the theory (we explain this additional structure in series of examples in Sect. 1.7.6). So we are actually going to think about them as $\mathbb{Z}$-graded categories (in fact, as dg-categories). But we should keep in mind that if we choose this additional structure in a different way, then a priori we should get different $\mathbb{Z}$-graded categories but with the same underlying $\mathbb{Z}_2$-graded categories.

[6]The nature of this additional structure will become more clear in Sect. 1.7.6.

Since a 3d $N = 4$ theory is supposed to have two topological twists which we call Coulomb and Higgs, we shall denote the corresponding categories of line operators by $\mathcal{C}_C, \mathcal{C}_H$. As was mentioned above, filling the circle with a disc should produce canonical objects $\mathcal{F}_C, \mathcal{F}_H$.

Remark for an Advanced Reader In principle in a true TQFT the category of line operators should be an E_2-category (cf. [35]). There is a closely related notion of factorizable category (in the D-module sense), a.k.a. chiral category, cf. [44]. In fact, the categories we are going to construct will be factorizable categories (and the canonical object, corresponding to the 2d disc will be a factorizable object). The fact that we get factorizable categories as opposed to E_2-categories is related to the warning in Sect. 1.7.7.

The relation between these structures and what we have discussed in the previous sections is that one should have

$$\mathrm{Ext}^*(\mathcal{F}_C, \mathcal{F}_C) = \mathbb{C}[\mathcal{M}_C] \tag{1.8}$$

and

$$\mathrm{Ext}^*(\mathcal{F}_H, \mathcal{F}_H) = \mathbb{C}[\mathcal{M}_H]. \tag{1.9}$$

Remark It can be shown that for any factorization category $\mathcal{C}$ and a factorization object $\mathcal{F}$ the algebra $\mathrm{Ext}^*(\mathcal{F}, \mathcal{F})$ is graded commutative.

When we need to emphasize dependence on a theory $\mathcal{T}$, we shall write $\mathcal{C}_C(\mathcal{T}), \mathcal{F}_C(\mathcal{T})$ etc. The mirror symmetry conjecture then says

Conjecture 1.21 The category $\mathcal{C}_C(\mathcal{T})$ is equivalent to $\mathcal{C}_H(\mathcal{T}^*)$ (and the same with C and H interchanged). Under this equivalence the object $\mathcal{F}_C(\mathcal{T})$ goes over to $\mathcal{F}_H(\mathcal{T}^*)$.

1.7.5 Generalities on D-Modules and de Rham Pre-stacks

In what follows we'll need to work with various categories of sheaves on spaces which are little more general than usual schemes or stacks. Namely, we need to discuss de Rham pre-stacks and various categories of sheaves related to them. Our main reference for the subject is [24].

Let S be a smooth scheme of finite type over $\mathbb{C}$. Then one can define certain pre-stack (i.e. a functor from $\mathbb{C}$-algebras to sets) S_{dR} which is called *the de Rham pre-stack of* S. Informally it is defined as the quotient of S by infinitesimal automorphisms. Moreover, this definition can be extended to all schemes, stacks or even dg-stacks of finite type over $\mathbb{C}$. A key property of S_{dR} is that the category of quasi-coherent sheaves on S_{dR} is the same as the category of D-modules on S.[7] In addition for a target stack $\mathcal{Y}$ one can consider the mapping space $\mathrm{Maps}(S_{dR}, \mathcal{Y})$. Here are two important examples:

(1) Let $\mathcal{Y} = \mathbb{A}^1$. Then $\mathrm{Maps}(S_{dR}, \mathcal{Y})$ is the de Rham cohomology of S considered as a dg-scheme.
(2) Let $\mathcal{Y} = \mathrm{pt}/G$ where G is an algebraic group. Then $\mathrm{Maps}(S_{dR}, \mathcal{Y})$ is the stack of G-local systems on S (i.e. the stack classifying G-bundles on S endowed with a flat connection).

In the sequel we'll need to apply these constructions to S being either the formal disc $\mathcal{D} = \mathrm{Spec}(\mathcal{O})$ or the punctured disc $\mathcal{D}^* = \mathrm{Spec}(\mathcal{K})$. This is not formally a special case of the above as some completion issues arise if one tries to spell out a careful definition. However, with some extra care all definitions can be extended to this case. This is done in [23].

1.7.6 Construction of the Categories in the Cotangent Case

It is expected that one can attach the above theories and categories to any symplectic dg-stack $\mathcal{X}$. It is now easy to spell out the additional structure on the theory that one needs in order to define the two topological twists in terms of the stack $\mathcal{X}$. Namely, one needs a $\mathbb{C}^\times$-action on $\mathcal{X}$ with respect to which the symplectic form ω has weight 2.

We shall actually assume that $\mathcal{X} = T^*\mathcal{Y}$ where $\mathcal{Y}$ is a smooth stack; in this case the above $\mathbb{C}^\times$-action is automatic (we can just use the square of the standard $\mathbb{C}^\times$-action on the cotangent fibers). We shall denote this theory by $\mathcal{T}(\mathcal{Y})$.

The following construction is due to T. Dimofte, D. Gaiotto, J. Hilburn and P. Yoo (private communication). Namely, let us set

$$\mathcal{C}_C = D\text{-mod}(\mathrm{Maps}(\mathcal{D}^*, \mathcal{Y})); \quad \mathcal{C}_H = \mathrm{QCoh}(\mathrm{Maps}(\mathcal{D}^*_{dR}, \mathcal{Y})). \tag{1.10}$$

Let us stress that both D-mod and QCoh mean the corresponding derived categories.

Let now $\pi_C\colon \mathrm{Maps}(\mathcal{D}_{dR}, \mathcal{Y}) \to \mathrm{Maps}(\mathcal{D}^*_{dR}, \mathcal{Y})$ be the natural map; similarly we define π_H. Then, we set

$$\mathcal{F}_H = (\pi_H)_*\mathcal{O}_{\mathrm{Maps}(\mathcal{D}_{dR}, \mathcal{Y})}; \quad \mathcal{F}_C = (\pi_C)_!\mathcal{O}_{\mathrm{Maps}(\mathcal{D}, \mathcal{Y})}. \tag{1.11}$$

[7] Because we plunge ourselves into world of derived algebraic geometry here, it doesn't make sense to talk about either quasi-coherent sheaves or D-modules as an abelian category: only the corresponding derived category makes sense.

1.7.7 A Very Important Warning

The above suggestion is probably only an approximation of a true statement. In fact, we believe that the suggestion is fine for $\mathcal{C}_C$; however, for $\mathcal{C}_H$ some modifications might be necessary. Let, for example (for simplicity), $\mathcal{Z}$ be a dg-stack of finite type over $\mathbb{C}$. Then following [2] in addition to the category $\mathrm{QCoh}(\mathcal{Z})$ one can also study the derived category $\mathrm{IndCoh}(\mathcal{Z})$ of ind-coherent sheaves on $\mathcal{Z}$. The two categories coincide when $\mathcal{Z}$ is a smooth classical (i.e. not dg) stack. But for more general $\mathcal{Z}$ these categories are different. This can be seen as follows: the compact objects of $\mathrm{IndCoh}(\mathcal{Z})$ are by definition finite complexes with coherent cohomology, while the compact objects of $\mathrm{QCoh}(\mathcal{Z})$ are finite perfect complexes. Moreover, assume that $\mathcal{Z}$ is locally a complete intersection. Then in [2] the authors define certain stack $\mathrm{Sing}(\mathcal{Z})$ endowed with a representable map $\mathrm{Sing}(\mathcal{Z}) \to \mathcal{Z}$, which is an isomorphism when $\mathcal{Z}$ is a smooth classical stack. Moreover, the fibers of this map are vector spaces; in particular, there is a natural $\mathbb{C}^\times$-action on the fibers whose stack of fixed points is naturally identified with $\mathcal{Z}$. Given a closed conical substack $\mathcal{W} \subset \mathrm{Sing}(\mathcal{Z})$ the authors in [2] define a category $\mathrm{IndCoh}_{\mathcal{W}}(\mathcal{Z})$ of ind-coherent sheaves with singular support in $\mathcal{W}$. These categories in some sense interpolate between $\mathrm{QCoh}(\mathcal{Z})$ and $\mathrm{IndCoh}(\mathcal{Z})$: namely, when $\mathcal{W} = \mathcal{Z}$ (the zero section of the morphism $\mathrm{Sing}(\mathcal{Z}) \to \mathcal{Z}$) we have $\mathrm{IndCoh}_{\mathcal{W}}(\mathcal{Z}) = \mathrm{QCoh}(\mathcal{Z})$, and when $\mathcal{W} = \mathrm{Sing}(\mathcal{Z})$ we have $\mathrm{IndCoh}_{\mathcal{W}}(\mathcal{Z}) = \mathrm{IndCoh}(\mathcal{Z})$.[8]

We think that suggestion (1.10) is only "the first approximation" to the right statement. More precisely, we believe that it is literally the right suggestion for the category $\mathcal{C}_C$, but for the category $\mathcal{C}_H$ one has to be more careful. We believe that the correct definition of the category $\mathcal{C}_H$ in the above context should actually be $\mathrm{IndCoh}_{\mathcal{W}}(\mathrm{Maps}(\mathcal{D}^*_{dR}, \mathcal{Y}))$ for a particular choice of $\mathcal{W}$ (very often $\mathcal{W}$ will actually be the zero section but probably not always); at this moment we don't know how to specify $\mathcal{W}$ in the above generality. The purpose of the rest of the section will be to explain some general picture, so in what follows we are going to ignore this subtlety, i.e. we shall proceed with the suggestion $\mathcal{C}_H = \mathrm{QCoh}(\mathrm{Maps}(\mathcal{D}^*_{dR}, \mathcal{Y}))$ as stated. But the reader should keep in mind that in certain cases it should be replaced by $\mathrm{IndCoh}_{\mathcal{W}}(\mathrm{Maps}(\mathcal{D}^*_{dR}, \mathcal{Y}))$ (this issue will become important when we formulate some rigorous conjectures (cf. for example the discussion before Conjecture 1.27).

1.7.8 Remarks About Rigorous Definitions

Since the above mapping spaces are often genuinely infinite-dimensional, we need to discuss why the above categories make sense. First, the category of D-modules on arbitrary pre-stack is discussed in [45]. In fact, in *loc. cit.* the author defines two

[8]The reader should be warned that although we have a natural functor $\mathrm{IndCoh}_{\mathcal{W}}(\mathcal{Z}) \to \mathrm{IndCoh}(\mathcal{Z})$, this functor is not fully faithful, so $\mathrm{IndCoh}_{\mathcal{W}}(\mathcal{Z})$ is not a full subcategory of $\mathrm{IndCoh}(\mathcal{Z})$.

versions of this category, which are denoted by $D^!$ and D^* (these two categories are dual to each other). For the purposes of these notes we need to work with D^*: for example since it is this category for which the functor of direct image is well-defined.

The category QCoh($\mathcal{Z}$) is well-defined for any pre-stack $\mathcal{Z}$; however in such generality it might be difficult to work with. However, we would like to note that $\mathrm{Maps}(\mathcal{D}^*_{dR}, \mathcal{Y})$ is typically a very manageable object. For example, when $\mathcal{Y}$ is a scheme of finite type over $\mathbb{C}$ it follows from Conjecture 1.23 below that $\mathrm{Maps}(\mathcal{D}^*_{dR}, \mathcal{Y})$ is a dg-scheme of finite type over $\mathbb{C}$, so QCoh is "classical" (modulo the fact that we have to work with commutative dg-algebras as opposed to usual commutative algebras). When $\mathcal{Y}$ is a stack, the definition is a bit less explicit; however, we claim that the definition is easy when $\mathcal{Y}$ is of the form $\mathcal{S}/G$ where $\mathcal{S}$ is an affine scheme and G is a reductive group. For example, when $\mathcal{Y} = \mathrm{pt}/G$ we have $\mathrm{Maps}(\mathcal{D}^*_{dR}, \mathcal{Y}) = \mathrm{LocSys}_G(\mathcal{D}^*)$: the stack of G-local systems (i.e. principal G-bundles with a connection on $\mathcal{D}^*$), and $\mathrm{QCoh}(\mathrm{LocSys}_G(\mathcal{D}^*))$ is a well-studied object in (local) geometric Langlands correspondence.

Here is another reason why we want $\mathcal{Y}$ to be of the above form. The map π_H is actually always a closed embedding, so we could write $(\pi_H)_!$ instead of $(\pi_H)_*$. On the other hand, the functor $(\pi_C)_!$ is a priori not well defined, at least it is not defined for an arbitrary morphism. However, it is well-defined if the morphism π_C is ind-proper. In what follows we shall always assume that the stack $\mathcal{Y}$ is such that it is the case. This condition is not always satisfied but it is also not super-restrictive as follows from the next exercise.

Exercise

(a) Show that π_C is a closed embedding if $\mathcal{Y}$ is a scheme.
(b) Show that if $\mathcal{Y} = \mathcal{S}/G$ where $\mathcal{S}$ is an affine scheme and G is a reductive algebraic group then the morphism π_C is ind-proper.
(c) Show that (b) might become false if we drop either the assumption that $\mathcal{S}$ is affine or the assumption that G is reductive.

We shall denote the corresponding categories (1.10) and objects (1.11) simply by $\mathcal{C}_C(\mathcal{Y})$, $\mathcal{F}_C(\mathcal{Y})$ etc. Note that these categories are $\mathbb{Z}$-graded. The above arguments then suggest the following

Conjecture 1.22 Let $\mathcal{Y}, \mathcal{Y}'$ be two stacks such that $T^*\mathcal{Y}$ is isomorphic to $T^*\mathcal{Y}'$ as a symplectic dg-stack. Then the corresponding $\mathbb{Z}_2$-graded versions of $\mathcal{C}_C(\mathcal{Y})$ and $\mathcal{C}_C(\mathcal{Y}')$ are equivalent as $\mathbb{Z}_2$-graded factorization categories; this equivalence sends $\mathcal{F}_C(\mathcal{Y})$ to $\mathcal{F}_C(\mathcal{Y}')$. Similar statement holds for $\mathcal{C}_H$.

1.7.9 Small Loops

In fact, one can demistify the category $\mathcal{C}_H(\mathcal{Y})$ a little bit which makes it quite computable. First of all, with the correct definition it is easy to see that $\mathrm{Maps}(\mathcal{D}_{dR}, \mathcal{Y})$

is equivalent to $\mathcal{Y}$ (for any $\mathcal{Y}$).[9] Now, given $\mathcal{Y}$ let us define another dg-stack $L\mathcal{Y}$ (we shall call it "small loops" into $\mathcal{Y}$) by setting

$$L\mathcal{Y} = \mathcal{Y} \underset{\mathcal{Y}\times\mathcal{Y}}{\times} \mathcal{Y},$$

where in the above equation both maps $\mathcal{Y} \to \mathcal{Y}\times\mathcal{Y}$ are equal to the diagonal map.[10] We have a natural map $\mathcal{Y} \to L\mathcal{Y}$.

Conjecture 1.23

1. Let $\mathcal{Y}$ be a scheme. Then $L\mathcal{Y}$ and $\mathrm{Maps}(\mathcal{D}^*_{dR}, \mathcal{Y})$ are isomorphic (and this isomorphism is compatible with the map from $\mathcal{Y} = \mathrm{Maps}(\mathcal{D}_{dR}, \mathcal{Y})$ into both).
2. Let $\mathcal{Y}$ be a stack. Then the formal neighbourhoods of $\mathcal{Y} = \mathrm{Maps}(\mathcal{D}_{dR}, \mathcal{Y})$ in $\mathrm{Maps}(\mathcal{D}^*_{dR}, \mathcal{Y})$ and in $L\mathcal{Y}$ are equivalent.

The proof of Conjecture 1.23 will be written in a different publication. In what follows we shall assume Conjecture 1.23.

1.7.10 Remark

If $\mathcal{Y}$ is a scheme then it is easy to see that both $\mathrm{Maps}(\mathcal{D}^*_{dR}, \mathcal{Y})$ and $L\mathcal{Y}$ are dg-extensions of $\mathcal{Y}$ (i.e. they are dg-schemes whose underlying classical scheme is $\mathcal{Y}$), so if the statement of Conjecture 1.23 holds on the level of formal neighbourhoods then in fact we have $L\mathcal{Y} = \mathrm{Maps}(\mathcal{D}^*_{dR}, \mathcal{Y})$. This is not the case for stacks. Namely, let G be a reductive algebraic group and let $\mathcal{Y} = \mathrm{pt}/G$. Then it is easy to see that $\mathrm{Maps}(\mathcal{D}^*_{dR}, \mathcal{Y})$ is the stack $\mathrm{LocSys}_G(\mathcal{D}^*)$ of G-local systems on $\mathcal{D}^*$ (i.e. principal G-bundles on $\mathcal{D}^*$ with a connection).

Exercise Show that for $\mathcal{Y} = \mathrm{pt}/G$ we have $L\mathcal{Y} = G/\mathrm{Ad}(G)$ (i.e. quotient of G by itself with respect to the adjoint action). Show that $G/\mathrm{Ad}(G)$ is not equivalent to $\mathrm{LocSys}_G(\mathcal{D}^*)$ but the formal neighbourhoods of pt/G in both are equivalent (the embedding of pt/G into $\mathrm{LocSys}_G(\mathcal{D}^*)$ corresponds to the trivial local system).

1.7.11 An Example

The significance of Conjecture 1.23 is that it allows to use the (very explicit) stack $L\mathcal{Y}$ in order to compute the Ext-algebra (1.9).

[9] Here we see that $\mathrm{Maps}(\mathcal{D}_{dR}, \mathcal{Y})$ should be defined with some extra care. Namely, if we just used the naive definition then the equivalence $\mathrm{Maps}(\mathcal{D}_{dR}, \mathcal{Y}) \simeq \mathcal{Y}$ would imply that $\mathcal{D}_{dR} = \mathrm{pt}$ which is far from being the case.

[10] Here we want to stress once again that all fibered products must be understood in the dg-sense!

Assume that $\mathcal{Y}$ is a smooth scheme which for simplicity we shall also assume to be affine. Then $L\mathcal{Y}$ is just the dg-scheme $\mathrm{Spec}(\mathrm{Sym}_{\mathcal{O}_\mathcal{Y}} T^*\mathcal{Y}[1])$. Since we have

$$\mathrm{Ext}^*_{\mathrm{Sym}_{\mathcal{O}_\mathcal{Y}} T^*\mathcal{Y}[1])}(\mathcal{O}_\mathcal{Y}, \mathcal{O}_\mathcal{Y}) = \mathrm{Sym}^{\bullet}_{\mathcal{O}_\mathcal{Y}}(T\mathcal{Y}[-2]),$$

we see that (with grading disregarded) $\mathbb{C}[\mathcal{M}_H] = \mathbb{C}[T^*\mathcal{Y}]$ which is what we should have in this case. In fact, if we want to rembember the grading we see that the homological grading on the RHS goes to grading coming from dilation of the cotangent fibers on the LHS. Recall that writing $\mathcal{X}$ as $T^*\mathcal{Y}$ is an additional structure which is precisely the one required in order to make all the categories $\mathbb{Z}$-graded (as opposed to $\mathbb{Z}_2$-graded; note also that the grading on $\mathrm{Sym}_{\mathcal{O}_\mathcal{Y}}(T\mathcal{Y}[-2])$ is even, so the corresponding $\mathbb{Z}_2$-grading is trivial).

Let us now compute $\mathbb{C}[\mathcal{M}_C]$ in this case. Since $\mathcal{Y}$ is an affine scheme, it follows that $\mathrm{Maps}(\mathcal{D}, \mathcal{Y})$ is a closed subscheme in the ind-scheme $\mathrm{Maps}(\mathcal{D}^*, \mathcal{Y})$, so $\mathrm{Ext}^*(\mathcal{F}_C, \mathcal{F}_C)$ is just equal to the de Rham cohomology of $\mathrm{Maps}(\mathcal{D}, \mathcal{Y})$. Since $\mathcal{Y}$ is smooth the (evaluation at $0 \in \mathcal{D}$) map $\mathrm{Maps}(\mathcal{D}, \mathcal{Y}) \to \mathcal{Y}$ is a fiber bundle whose fibers are infinite-dimensional affine spaces. Thus it induces an isomorphism on de Rham cohomology. Hence we get $\mathbb{C}[\mathcal{M}_C] = H^*(\mathcal{Y}, \mathbb{C}) = H^*(T^*\mathcal{Y}, \mathbb{C})$. So if $\mathcal{Y}$ is connected, we see that $\mathcal{M}_C$ is a dg-extension of pt; moreover, if $\mathcal{Y}$ is a vector space, that $\mathcal{M}_C = \mathrm{pt}$ even as dg-schemes (as was promised in Sect. 1.4.4).

1.7.12 Gauge Theory

Consider now the example when $\mathcal{Y} = \mathbf{N}/G$, where G is a connected reductive group and $\mathbf{N}$ is a representation of G.

Exercise Show that in this case the LHS of (1.8) is literally the same as $H^{G_\mathcal{O}}_\bullet(\mathcal{R}_{G,\mathbf{N}})$.

So, we see that our categorical point of view recovers the definition of the Coulomb branch we gave before. Let us look at the Higgs branch. According to Conjecture 1.23 we need to understand the dg-stack

$$L(\mathbf{N}/G) = \mathbf{N}/G \underset{\mathbf{N}/G \times \mathbf{N}/G}{\times} \mathbf{N}/G. \tag{1.12}$$

Let us actually first assume that $\mathbf{N}$ is any smooth variety with a G-action. Then it is easy to see that (1.12) is a dg-stack which admits the following description. The action of G on $\mathbf{N}$ defines a natural map of locally free $\mathcal{O}_\mathbf{N}$-modules

$$\mathfrak{g} \otimes \mathcal{O}_\mathbf{N} \to T\mathbf{N}.$$

Consider the dual map

$$T^*\mathbf{N} \to \mathfrak{g}^* \otimes \mathcal{O}_\mathbf{N}$$

and let us regard it as two step complex of coherent sheaves on $\mathbf{N}$ where $T^*\mathbf{N}$ lives in degree -1 and $\mathfrak{g}^* \otimes \mathcal{O}_\mathbf{N}$ lives in degree 0. Let us denote this complex by $K^\bullet$. Then $\mathrm{Sym}_{\mathcal{O}_\mathbf{N}}(K^\bullet)$ is a quasi-coherent dg-algebra on $\mathbf{N}$.

Exercise Show that the formal neighbourhood of $\mathbf{N}/G$ in $L(\mathbf{N}/G)$ is equivalent to the formal neighbourhood of $\mathbf{N}/G$ in $\mathrm{Spec}(\mathrm{Sym}_{\mathcal{O}_\mathbf{N}}(K^\bullet))/G$ (note that when G is trivial we just recover $\mathrm{Spec}(\mathrm{Sym}(T^*\mathbf{N})[1]))$ as in the previous subsection).

It now follows that the LHS of (1.9) in our case becomes equal to the G-invariant part of

$$\mathrm{Ext}^*_{\mathrm{Sym}_{\mathcal{O}_\mathbf{N}}(K^\bullet)}(\mathcal{O}_\mathbf{N}, \mathcal{O}_\mathbf{N}). \tag{1.13}$$

Assume now for simplicity that $\mathbf{N}$ is affine. Then it is easy to see that (1.13) is equal to the cohomology of $\mathrm{Sym}_{\mathcal{O}_\mathbf{N}}((K^\bullet)^*[-1])$.

Exercise Show that as a $\mathbb{Z}_2$-graded algebra $\mathrm{Sym}_{\mathcal{O}_\mathbf{N}}((K^\bullet)^*[-1])$ is quasi-isomorphic to the algebra of functions on the dg-scheme $\mu^{-1}(0)$ where $\mu : T^*\mathbf{N} \to \mathfrak{g}^*$ is the moment map.

The exercise implies that the LHS of (1.9) is isomorphic to the algebra of functions on the dg-stack $\mu^{-1}(0)/G$.

1.7.13 Mirror Symmetry in the Toric Case

Let us assume that we are in the situation of Sect. 1.4.6. We set $\mathcal{Y} = \mathbb{C}^n/T$, $\mathcal{Y}^* = \mathbb{C}^n/T_F^\vee$. Combining (1.10) and (1.11) with Conjecture 1.21 we already obtain a bunch of non-trivial statements. Namely, we arrive at the following

Conjecture 1.24 For the above choice of $\mathcal{Y}$ and $\mathcal{Y}^*$ we have equivalences of (factorization) categories

$$D\text{-mod}(\mathrm{Maps}(\mathcal{D}^*, \mathcal{Y})) \simeq \mathrm{QCoh}(\mathrm{Maps}(\mathcal{D}^*_{dR}, \mathcal{Y}^*))$$

$$D\text{-mod}(\mathrm{Maps}(\mathcal{D}^*, \mathcal{Y}^*)) \simeq \mathrm{QCoh}(\mathrm{Maps}(\mathcal{D}^*_{dR}, \mathcal{Y})).$$

A proof of this conjecture is the subject of a current work in progress of the first named author with Dennis Gaitsgory. Let us discuss the simplest example (which is already quite non-trivial).

Let us take $n = 1$ and let T be trivial. In other words we get $\mathcal{Y} = \mathbb{A}^1$. Then $\mathcal{Y}^* = \mathbb{A}^1/\mathbb{G}_m$. So, let us look closely at what Conjecture 1.24 says in this case.

First, $\mathrm{Maps}(\mathcal{D}^*_{dR}, \mathcal{Y}) = \mathrm{Maps}(\mathcal{D}^*_{dR}, \mathbb{A}^1) = H^*_{dR}(\mathcal{D}^*) = \mathbb{A}^1 \times \mathbb{A}^1[-1]$. In other words, $\mathrm{Maps}(\mathcal{D}^*_{dR}, \mathcal{Y}) = \mathrm{Spec}(\mathbb{C}[x, \varepsilon])$ where $\deg(x) = 0, \deg(\varepsilon) = -1$ (we consider it as a dg-algebra with trivial differential). The category $\mathcal{C}_H(\mathcal{Y})$ is then just the derived category of dg-modules over this algebra. More precisely, it is the QCoh version of this derived category—we again refer the reader to Chapter II of [24]. The object $\mathcal{F}_H$ corresponds to the dg-module $\mathbb{C}[x]$ (on which ε acts trivially) in degree 0.

We claim that in this case the category $\mathrm{QCoh}(\mathrm{Maps}(\mathcal{D}^*_{dR}, \mathbb{A}^1))$ is equivalent to $D\text{-mod}(\mathrm{Maps}(\mathcal{D}^*, \mathbb{A}^1/\mathbb{G}_m))$ even as a $\mathbb{Z}$-graded category. We are not in a position to give a rigorous proof here, since for this we'll need to spell out careful definitions of both categories, and that goes beyond the scope of these notes. Let us give some examples of objects which go to one another under the above equivalence. First, the object $\mathcal{F}_C \in D\text{-mod}(\mathrm{Maps}(\mathcal{D}^*, \mathbb{A}^1/\mathbb{G}_m))$ is described as follows. Let $i_n : \mathcal{O} \to \mathcal{K}$ be the embedding which sends f to $z^n f$ (here $n \in \mathbb{Z}$). Then we have

$$\mathcal{F}_C(\mathbb{A}^1/\mathbb{G}_m) = \bigoplus_{n \in \mathbb{Z}} (i_n)_* \mathcal{O}.$$

Warning To understand this object carefully one really needs to spell out the definition. Let us mention the problem one has to fight with. It is intuitively clear that we have a $\mathbb{Z}$-action on $\mathcal{K}$ such that $n \in \mathbb{Z}$ sends $f(z)$ to $z^n f(z)$. On the other hand, assume that $n \geq 0$. Then $z^n \mathcal{O}$ has codimension n in $\mathcal{O}$ (although one is obtained from the other by means of the $\mathbb{Z}$-action). This problem is in fact not as serious as it might seem at the first glance—it just shows that the actual definition of D-modules on $\mathcal{K}$ (or even on $\mathcal{O}$) must take into account certain homological shifts.

Having the above warning in mind, it is easy to see that $\mathrm{Ext}^*(\mathcal{F}_C, \mathcal{F}_C) = \mathbb{C}[x, y]$ where $\deg(x) = 0, \deg(y) = 2$. On the other hand, we also have

$$\mathrm{Ext}^*_{\mathbb{C}[x,\varepsilon]}(\mathbb{C}[x], \mathbb{C}[x]) = \mathbb{C}[x, y],$$

which matches our expectations.

Here is another example. Consider the module $\mathbb{C}$ over $\mathbb{C}[x, \varepsilon]$ (i.e. we think of it as a dg-module concentrated in degree 0, on which x acts by 1 and ε acts by 0). Then under the above equivalence it goes to the D-module δ of delta-functions at $0 \in \mathcal{K}$ (considered as a $\mathcal{K}^*$-equivariant D-module). Note that the $\mathcal{K}^*$-equivariant Ext from δ to itself is the same as $H^*_{\mathcal{K}^*}(\mathrm{pt}, \mathbb{C})$. Now, homotopically $\mathcal{K}^*$ is equivalent to $\mathbb{C}^\times \times \mathbb{Z}$ and we have

$$H^*_{\mathbb{C}^\times \times \mathbb{Z}}(\mathrm{pt}, \mathbb{C}) = \mathbb{C}[y, \theta] \quad \text{where } \deg(y) = 2, \deg(\theta) = 1.$$

On the other hand the same (dg) algebra $\mathbb{C}[y, \theta]$ is equal to $\mathrm{Ext}^*_{\mathbb{C}[x,\varepsilon]}(\mathbb{C}, \mathbb{C})$.

1.7.14 The Theory $\mathcal{T}[G]$

Here is another expectation. Let $D(\mathrm{Gr}_G)^{\mathrm{Hecke}}$ denote the derived category of Hecke eigen-modules on Gr_G, i.e. D-modules which are also right modules for the algebra $\mathcal{A}_R$.

Conjecture 1.25 The category $\mathcal{C}_C(\mathcal{T}[G])$ is the category $D(\mathrm{Gr}_G)^{\mathrm{Hecke}}$ and $\mathcal{F}_C = \mathcal{A}_R$.

Let us combine it with (1.10). In the case when $G = \mathrm{GL}(n)$ the theory $\mathcal{T}[G]$ does in fact come from a smooth stack $\mathcal{Y}$; here

$$\mathcal{Y} = (\prod_{i=1}^{n-1} \mathrm{Hom}(\mathbb{C}^i, \mathbb{C}^{i+1}))/\prod_{i=1}^{n-1} \mathrm{GL}(i) \tag{1.14}$$

(note that $\mathcal{Y}$ still has an action of $\mathrm{GL}(n)$). So, from (1.10) we get another construction of $\mathcal{C}_C$ which should be equivalent to the one from Conjecture 1.25.

It is in fact easy to construct a functor in one direction. Namely, let $\mathcal{C}$ be a category with a D-module action of some group G; let also $\mathcal{F}$ be a $G_{\mathcal{O}}$-equivariant object. Then $\mathcal{F}$ defines a functor $\mathcal{C} \to D\text{-mod}(\mathrm{Gr}_G)$. Moreover, this functor sends $\mathcal{F}$ to a ring object $\mathcal{A}_{\mathcal{F}}$ and the above functor can be upgraded to a functor from $\mathcal{C}$ to $\mathcal{A}_{\mathcal{F}}$-modules in $D\text{-mod}(\mathrm{Gr}_G)$. Namely, this functor sends every $\mathcal{G}$ to the D-module on Gr_G whose !-stalk at some g is equal to $\mathrm{RHom}(\mathcal{F}^g, \mathcal{G})$. In our case we take $\mathcal{C}$ to be the category of D-modules on $\mathrm{Maps}(\mathcal{D}^*, \mathcal{Y})$ and take $\mathcal{F} = \mathcal{F}_C$. Then the above functor sends $\mathcal{F}$ to $\mathcal{A}_R$ (this is essentially proved in [15]).

Note that for $G = \mathrm{GL}(n)$ the theory $\mathcal{T}[G]$ is supposed to be self-dual (with respect to mirror symmetry procedure). Hence it follows that in this case the category $\mathcal{C}_C$ should be equivalent to $\mathcal{C}_H$. Therefore, it is natural to expect that the category $\mathrm{QCoh}(\mathrm{Maps}(\mathcal{D}^*_{dR}, \mathcal{Y}))$ is equivalent to $D(\mathrm{Gr}_{\mathrm{GL}(n)})^{\mathrm{Hecke}}$. However, we expect that it is actually wrong as stated—the reason is the warning from Sect. 1.7.7. However, we do believe in the following

Conjecture 1.26 Let $\mathcal{Y}$ be as in (1.14). Then the category $\mathrm{IndCoh}(\mathrm{Maps}(\mathcal{D}^*_{dR}, \mathcal{Y}))$ is equivalent to $D(\mathrm{Gr}_{\mathrm{GL}(n)})^{\mathrm{Hecke}}$.

Here is (an equivalent) variant of this conjecture. Note that the action of the group $\mathrm{GL}(n)$ on $\mathcal{Y}$ gives rise to an action of the same group on $\mathrm{Maps}(\mathcal{D}^*_{dR}, \mathcal{Y})$. Hence we can consider the category $\mathrm{QCoh}(\mathrm{Maps}(\mathcal{D}^*_{dR}, \mathcal{Y})/\mathrm{GL}(n))$. This category admits a natural action of the tensor category $\mathrm{Rep}(\mathrm{GL}(n))$. Note that the geometric Satake equivalence also gives rise to an action of $\mathrm{Rep}(\mathrm{GL}(n))$ on $D\text{-mod}(\mathrm{Gr}_{\mathrm{GL}(n)})$ (the action is by convolution with spherical D-modules on the right).

Conjecture 1.27 The categories $\mathrm{IndCoh}(\mathrm{Maps}(\mathcal{D}^*_{dR}, \mathcal{Y})/\mathrm{GL}(n))$ and D-mod $(\mathrm{Gr}_{\mathrm{GL}(n)})$ are equivalent as module categories over $\mathrm{Rep}(\mathrm{GL}(n))$.

In the paper [11] we prove a weaker version of Conjecture 1.27 for GL(2) (in particular, the version of Conjecture 1.27 proved in [11] is sufficient in order to explain why we need IndCoh and not QCoh in the formulation).

1.7.15 G-Symmetry and Gauging

Let us now address the following question. Let $\mathcal{T}$ be a theory acted on by a (reductive) algebraic group G. What kind of structures does this action imply in terms of the categories $\mathcal{C}_H$, $\mathcal{C}_C$?

To answer this question, we need to recall two general notions. First, given a category $\mathcal{C}$ and a group ind-scheme $\mathcal{G}$ there is a notion *strong* or *infinitesimally trivial* $\mathcal{G}$-action on $\mathcal{C}$ (cf. [22]). The main example of such an action is as follows: given a pre-stack $\mathcal{S}$ with a $\mathcal{G}$-action, the group $\mathcal{G}$ acts strongly on the (derived) category of D-modules on $\mathcal{S}$. If one replaces D-modules by quasi-coherent sheaves, one gets the notion of *weak* G-action on a category $\mathcal{C}$. Given a category $\mathcal{C}$ with a strong $\mathcal{G}$-action one can define the category of *strongly equivariant objects in* $\mathcal{C}$ (cf. [22, page 4]); we shall denote this category by $\mathcal{C}^{\mathcal{G}}$.

On the other hand, for a stack $\mathcal{Z}$ and a (dg-)category $\mathcal{C}$ there is a notion of "$\mathcal{C}$ living over $\mathcal{Z}$" (cf. [21]). This simply means that the category QCoh($\mathcal{Z}$) (which is a tensor category) acts on $\mathcal{C}$. Given a geometric point z of $\mathcal{Z}$ we can consider the fiber $\mathcal{C}_z$ of $\mathcal{C}$ at z. This category always has a weak action of the group Aut_z of automorphisms of the point z.

Now we can formulate an (approximate) answer to the above question. Namely, we expect that a G-action on $\mathcal{T}$ should produce the following structures:

(1) A category $\mathcal{C}_H(G, \mathcal{T})$ which lives over $\mathrm{LocSys}_G(\mathcal{D}^*)$ endowed with an equivalence

$$\mathcal{C}_H(G, \mathcal{T})_{\mathbf{Triv}} \simeq \mathcal{C}_H(T).^{11}$$

Here **Triv** stands for the trivial local system.

(2) A strong $G(\mathcal{K}) = \mathrm{Maps}(\mathcal{D}^*, G)$-action on the category $\mathcal{C}_C(\mathcal{T})$.

Note that G is the group of automorphisms of the trivial local system. Hence (1) implies that a G-action on $\mathcal{T}$ yields a weak action of G on $\mathcal{C}_H(\mathcal{T})$.

Exercise Show that this action extends to a weak action of LG (which is a dg-extension of G) on $\mathcal{C}$.

The reader must be warned that a weak action of G or even of LG on $\mathcal{C}_H(\mathcal{T})$ is a very small amount of data: for example, it is not sufficient in order to reconstruct $\mathcal{C}_H(G, \mathcal{T})$.

[11] Again, the reader should keep in mind Sect. 1.7.7.

Recall now that if a group G acts on a theory $\mathcal{T}$ then we can form the corresponding gauge theory $\mathcal{T}/G$. Then we expect that

$$\mathcal{C}_H(\mathcal{T}/G) = \mathcal{C}(G, \mathcal{T}); \quad \mathcal{C}_C(\mathcal{T}/G) = \mathcal{C}_C(\mathcal{T})^{G(\mathcal{K})}. \tag{1.15}$$

Let us now go back to the case $\mathcal{T} = \mathcal{T}(\mathcal{Y})$. In this case an action of G on $\mathcal{Y}$ yields an action of G on $\mathcal{T}(\mathcal{Y})$. In this case we expect that $\mathcal{T}(\mathcal{Y})/G = \mathcal{T}(\mathcal{Y}/G)$. Let us discuss the compatibility of this statement with above categorical structures. First, an action of G on $\mathcal{Y}$ gives rise to an action of $G(\mathcal{K})$ on $\mathrm{Maps}(\mathcal{D}^*, \mathcal{Y})$, hence a strong action on D-mod$(\mathrm{Maps}(\mathcal{D}^*, \mathcal{Y}))$. Moreover, $D\text{-mod}(\mathrm{Maps}(\mathcal{D}^*, \mathcal{Y}))^{G(\mathcal{K})} = D\text{-mod}(\mathrm{Maps}(\mathcal{D}^*, \mathcal{Y}/G))$ which is compatible with the second equation of (1.15). On the other hand, it is easy to see that the category $\mathrm{QCoh}(\mathrm{Maps}(\mathcal{D}^*_{dR}, \mathcal{Y}/G))$ lives over $\mathrm{QCoh}(\mathrm{LocSys}_G(\mathcal{D}^*))$ and its fiber over **Triv** is $\mathrm{QCoh}(\mathrm{Maps}(\mathcal{D}^*_{dR}, \mathcal{Y}))$ which is compatible with the first equation of (1.15).

1.7.16 S-Duality and Local Geometric Langlands

This subsection is a somewhat side topic: here we would like to mention a possible connection of the above discussion with (conjectural) local geometric Langlands correspondence. A reader who is not interested in the subject is welcome to skip this subsection.

The local geometric Langlands duality predicts the existence of an equivalence $\mathbf{L}_G$ between the (∞-)category of (dg-)categories with strong $G(\mathcal{K})$-action and the (∞-)category of (dg-)categories over $\mathrm{QCoh}(\mathrm{LocSys}_{G^\vee}(\mathcal{D}^*))$ (as was already mentioned earlier in these notes we are going to ignore higher categorical structures, which are in fact necessary in order to discuss these things rigorously).[12]

Let us now recall that given a theory $\mathcal{T}$ with a G-action one expects the existence of the S-dual theory $\mathcal{T}^\vee$ with a $G^\vee$-action. Thus we see that we get a category $\mathcal{C}_C(\mathcal{T}^\vee)$ with a strong $G^\vee(\mathcal{K})$-action and a category $\mathcal{C}_H(\mathcal{T}^\vee/G^\vee)$ which lives over $\mathrm{LocSys}_{G^\vee}(\mathcal{D}^*)$.

Conjecture 1.28 We have natural equivalences

$$\mathbf{L}_G(\mathcal{C}_C(\mathcal{T})) \simeq \mathcal{C}_H(\mathcal{T}^\vee/G^\vee); \quad \mathbf{L}_{G^\vee}(\mathcal{C}_C(\mathcal{T}^\vee)) \simeq \mathcal{C}_H(\mathcal{T}/G).$$

Recall now formula (1.2):

$$\mathcal{T}^\vee = ((\mathcal{T} \times \mathcal{T}[G])/G)^*.$$

[12] It is known that this is only an approximate conjecture. The correct conjecture (due to A. Arinkin) requires a (rather tricky) modification of the notion category over $\mathrm{LocSys}_G(\mathcal{D}^*)$ (which again has to do with the difference between QCoh and IndCoh).

In particular, we can apply it to $\mathcal{T}$ being the trivial theory; in this case we get that the group $G^\vee$ should act on the theory $(\mathcal{T}[G]/G)^*$. Let $\mathcal{C}_G = \mathcal{C}_C((\mathcal{T}[G]/G)^*)$. Then this category should have a strong action of $G^\vee(\mathcal{K})$. On the other hand, $\mathcal{C}_G = \mathcal{C}_H(\mathcal{T}[G]/G)$, so in addition it should live over $\mathrm{LocSys}_G(\mathcal{D}^*)$ (these two structures should commute in the obvious way). We expect that $\mathcal{C}_G$ is the *universal Langlands category* for $G^\vee$, i.e. that for any other category $\mathcal{C}$ with $G^\vee(\mathcal{K})$-action we have[13]

$$\mathbf{L}_{G^\vee}(\mathcal{C}) = \mathcal{C} \underset{G^\vee(\mathcal{K})}{\otimes} \mathcal{C}_G.$$

In particular, if $G = \mathrm{GL}(n)$ then we see that the universal Langlands category $\mathcal{C}_{\mathrm{GL}(n)}$ is expected to be equivalent to $\mathrm{QCoh}(\mathrm{Maps}(\mathcal{D}^*_{dR}, \mathcal{Y}/G))$ where $\mathcal{Y}$ is given by (1.14). Note that in this realization the fact that this category lives over $\mathrm{LocSys}_{\mathrm{GL}(n)}(\mathcal{D}^*)$ is clear, but the action of $\mathrm{GL}(n, \mathcal{K})$ is absolutely not obvious: we don't know how to construct it.

Again, it must be noted that the notion of universal Langlands category is not precise since as was mentioned above the correct formulation of the local geometric Langlands conjecture involves a modification of the notion of category over $\mathrm{LocSys}_G(\mathcal{D}^*)$. But at least we believe that the above description of the universal Langlands category is true as stated over the locus of irreducible local systems.

1.7.17 *Quantization*

Let us now discuss the categorical structures which give rise to the quantizations (and thus to Poisson structures) of the algebras $\mathbb{C}[\mathcal{M}_H]$ and $\mathbb{C}[\mathcal{M}_C]$. Let us first look at the latter one. The space $\mathrm{Maps}(\mathcal{D}^*, \mathcal{Y})$ has a natural action of the multiplicative group $\mathbb{G}_m$ (which acts on $\mathcal{D}^*$ by loop rotation). Thus the category $\mathcal{C}_C(\mathcal{Y})$ admits a natural deformation: the category $D\text{-mod}_{\mathbb{G}_m}(\mathrm{Maps}(\mathcal{D}^*, \mathcal{Y}))$ of $\mathbb{G}_m$-equivariant D-modules. The object $\mathcal{F}_C$ deforms naturally to an object of $D\text{-mod}_{\mathbb{G}_m}(\mathrm{Maps}(\mathcal{D}^*, \mathcal{Y}))$ and thus we can set

$$\mathbb{C}_\hbar(\mathcal{M}_C) = \mathrm{Ext}^*_{D\text{-mod}_{\mathbb{G}_m}(\mathrm{Maps}(\mathcal{D}^*, \mathcal{Y}))}(\mathcal{F}_C, \mathcal{F}_C).$$

Here $\hbar$ as before is a generator of $H^*_{\mathbb{G}_m}(\mathrm{pt}, \mathbb{C})$.

What about the quantization of $\mathcal{M}_H$? As before we need to look for a one-parameter deformation of the pair $(\mathcal{C}_H, \mathcal{F}_H)$. Here again the action of the multiplicative group $\mathbb{G}_m$ on $\mathcal{D}$ and on $\mathcal{D}^*$ gives rise to an action of $\mathbb{G}_m$ on the category $\mathcal{C}_H(\mathcal{Y})$; we claim that this action is strong (this is related to the fact that we work with maps from $\mathcal{D}^*_{dR}$ rather than with maps from $\mathcal{D}^*$). Thus it makes sense to consider

[13] Such a tensor product does make sense as long as we live in the world of dg-categories.

the category of strongly equivariant objects in $\mathcal{C}_H(\mathcal{Y})$ (cf. again [22, page 4]). The Ext-algebra of (the natural analog of) the object $\mathcal{F}_H$ in this category is again a non-commutative algebra over $\mathbb{C}[\hbar]$ which is a quantization of $\mathbb{C}[\mathcal{M}_H]$.

Note that since in both cases we use the action of the multiplicative group on $\mathcal{D}^*$, it follows that the deformed categories are no longer factorisation categories, so the corresponding Ext-algebras no longer have factorisation structure. This is why they have a chance to become non-commutative (at least the Remark after (1.9) does not apply here).

1.7.18 Holomorphic-Topological Twist

We have learned the main ideas of this subsection from K. Costello. So far we discussed the two topological twists of a given theory completely independently of each other. However, in fact in physics both the C-twist and the H-twist appear as one-parametric families of equivalent twists. In addition, both families have *the same* limiting point, where the theory is no longer topological (it becomes *holomorphic-topological*, cf. [1]; roughly speaking it means that, for example, for a 3-manifold M the partition function $Z(M)$ is well-defined if one fixes some additional structure on M which locally makes it look like a product of a complex curve Σ and a 1-manifold I). The category of line operators in the holomorphic-topological theory is still well-defined. As a result we come to the following conclusion:

Conclusion There should exists a factorisation category $\mathcal{C}$ with an object $\mathcal{F}$ and two $\mathbb{Z}$-gradings such that

(1) The two $\mathbb{Z}$-gradings yield the same $\mathbb{Z}_2$-grading.
(2) The pair $(\mathcal{C}_C, \mathcal{F}_C)$ is a deformation of the pair $(\mathcal{C}, \mathcal{F})$. This deformation preserves the 1st grading on $\mathcal{C}$.
(3) The pair $(\mathcal{C}_H, \mathcal{F}_H)$ is a deformation of the pair $(\mathcal{C}, \mathcal{F})$. This deformation preserves the 2nd grading on $\mathcal{C}$.

Let us describe a suggestion for the category $\mathcal{C}(\mathcal{Y})$ and the object $\mathcal{F}(\mathcal{Y})$. We would like to set

$$\mathcal{C}(\mathcal{Y}) = \mathrm{QCoh}(T^* \mathrm{Maps}(\mathcal{D}^*, \mathcal{Y})).$$

Here there are some technical problems: the stack $T^* \mathrm{Maps}(\mathcal{D}^*, \mathcal{Y})$ is very essentially infinite-dimensional, so studying quasi-coherent sheaves on it is more difficult than before. Let us assume that it is possible though and let us discuss (1)–(3) in this case. First, we need two gradings. The first grading is simply the homological grading on QCoh. The second grading is the combination of the homological grading and the grading coming from $\mathbb{C}^\times$-action on $T^* \mathrm{Maps}(\mathcal{D}^*, \mathcal{Y})$ (which dilates the cotangent fibers) multiplied by two (so the two grading manifestly yield the same $\mathbb{Z}_2$-grading).

Now the category of D-modules on $\text{Maps}(\mathcal{D}^*, \mathcal{Y})$ is clearly a deformation of $\text{QCoh}(T^* \text{Maps}(\mathcal{D}^*, \mathcal{Y}))$. On the other hand, it is less clear how to deform the category $\text{QCoh}(T^* \text{Maps}(\mathcal{D}^*, \mathcal{Y}))$ to $\text{QCoh}(\text{Maps}(\mathcal{D}^*_{dR}, \mathcal{Y}))$. We plan to address these issues in a future publication.

Acknowledgements We are greatly indebted to our coauthor H. Nakajima who taught us everything we know about Coulomb branches of 3d $N = 4$ gauge theories and to the organizers of the CIME summer school "Geometric Representation Theory and Gauge Theory" in June 2018, for which these notes were written. In addition we would like to thank T. Dimofte, D. Gaiotto, J. Hilburn and P. Yoo for their patient explanations of various things (in particular, as was mentioned above the main idea of Sect. 1.7 is due to them). Also, we are very grateful to R. Bezrukavnikov, K. Costello, D. Gaitsgory and S. Raskin for their help with many questions which arose during the preparation of these notes. The research of M.F. was supported by the grant RSF 19-11-00056.

References

1. M. Aganagic, K. Costello, J. McNamara, C. Vafa, Topological Chern-Simons/matter theories. arxiv:1706.09977
2. D. Arinkin, D. Gaitsgory, Singular support of coherent sheaves and the geometric Langlands conjecture. Selecta Math. (N.S.) **21**(1), 1–199 (2015)
3. S. Arkhipov, R. Bezrukavnikov, V. Ginzburg, Quantum groups, the loop Grassmannian, and the Springer resolution. J. Am. Math. Soc. **17**(3), 595–678 (2004)
4. M.F. Atiyah, Topological quantum field theories. Inst. Hautes Études Sci. Publ. Math. **68**, 175–186 (1988)
5. A. Beilinson, V. Drinfeld, Quantization of Hitchin's integrable system and Hecke eigensheaves (2000). http://www.math.uchicago.edu/mitya/langlands.html
6. R. Bezrukavnikov, M. Finkelberg, Equivariant Satake category and Kostant-Whittaker reduction. Moscow Math. J. **8**(1), 39–72 (2008)
7. R. Bielawski, A.S. Dancer, The geometry and topology of toric hyperkähler manifolds. Commun. Anal. Geom. **8**(4), 727–760 (2000)
8. T. Braden, Hyperbolic localization of intersection cohomology. Transform. Groups **8**, 209–216 (2003)
9. T. Braden, N. Proudfoot, B. Webster, Quantizations of conical symplectic resolutions I: local and global structure. Astérisque **384**, 1–73 (2016)
10. T. Braden, A. Licata, N. Proudfoot, B. Webster, Quantizations of conical symplectic resolutions II: category O and symplectic duality. Astérisque **384**, 75–179 (2016)
11. A. Braverman, M. Finkelberg, A quasi-coherent description of the category of D-mod$(Gr_{GL(n)})$. arXiv:1809.10774
12. A. Braverman, M. Finkelberg, D. Gaitsgory, Uhlenbeck spaces via affine Lie algebras. Progress in Mathematics, vol. 244 (Birkhäuser, Boston, 2006), pp. 17–135. Erratum. arxiv:0301176v4
13. A. Braverman, M. Finkelberg, H. Nakajima, Towards a mathematical definition of 3-dimensional $\mathcal{N} = 4$ gauge theories, II. Adv. Theor. Math. Phys. **22**(5), 1017–1147 (2018)
14. A. Braverman, M. Finkelberg, H. Nakajima, Coulomb branches of $3d$ $\mathcal{N} = 4$ quiver gauge theories and slices in the affine Grassmannian (with appendices by Alexander Braverman, Michael Finkelberg, Joel Kamnitzer, Ryosuke Kodera, Hiraku Nakajima, Ben Webster, and Alex Weekes). Adv. Theor. Math. Phys. **23**(1), 75–166 (2019)
15. A. Braverman, M. Finkelberg, H. Nakajima, Ring objects in the equivariant derived Satake category arising from Coulomb branches (with appendix by Gus Lonergan). Adv. Theor. Math. Phys. **23**(2), 253–344 (2019)

16. A. Braverman, M. Finkelberg, H. Nakajima, Line bundles over Coulomb branches. arXiv:1805.11826
17. S. Cremonesi, A. Hanany, A. Zaffaroni, Monopole operators and Hilbert series of Coulomb branches of $3d$ $\mathcal{N} = 4$ gauge theories. J. High Energy Phys. **1401**, 5 (2014)
18. V. Drinfeld, D. Gaitsgory, On a theorem of Braden. Transform. Groups **19**, 313–358 (2014)
19. M. Finkelberg, I. Mirković, Semi-infinite flags. I. Case of global curve $\mathbb{P}^1$. American Mathematical Society Translations Series 2, vol. 194 (American Mathematical Society, Providence, RI, 1999), pp. 81–112
20. D. Gaiotto, E. Witten, S-duality of boundary conditions in $\mathcal{N} = 4$ super Yang-Mills theory. Adv. Theor. Math. Phys. **13**(3), 721–896 (2009)
21. D. Gaitsgory, The notion of category over an algebraic stack. arXiv:math/0507192
22. D. Gaitsgory, Groups acting on categories. http://www.math.harvard.edu/~gaitsgde/grad_2009/SeminarNotes/April6(GrpActCat).pdf
23. D. Gaitsgory, Day VI, Talk 1. OPERS, Hebrew University school on geometric Langlands (2014)
24. D. Gaitsgory, N. Rozenblyum, A Study in Derived Algebraic Geometry. Mathematical Surveys and Monographs, vol. 221 (American Mathematical Society, Providence, RI, 2017)
25. V. Ginzburg, Perverse sheaves on a loop group and Langlands' duality. arXiv:alg-geom/9511007
26. V. Ginsburg, Perverse sheaves and $\mathbb{C}^*$-actions. J. Am. Math. Soc. **4**(3), 483–490 (1991)
27. V. Ginzburg, D. Kazhdan, Construction of symplectic varieties arising in 'Sicilian theories' (in preparation)
28. V. Ginzburg, S. Riche, Differential operators on G/U and the affine Grassmannian. J. Inst. Math. Jussieu **14**(3), 493–575 (2015)
29. A. Grothendieck, Sur la classification des fibrés holomorphes sur la sphère de Riemann. Am. J. Math. **79**, 121–138 (1957)
30. M. Kashiwara, The flag manifold of Kac-Moody Lie algebra, *in Algebraic Analysis, Geometry and Number Theory* (Baltimore, 1988) (Johns Hopkins University Press, Baltimore, MD, 1989) pp. 161–190
31. M. Kashiwara, T. Tanisaki, Kazhdan-Lusztig conjecture for affine Lie algebras with negative level. Duke Math. J. **77**, 21–62 (1995)
32. J. Kamnitzer, B. Webster, A. Weekes, O. Yacobi, Yangians and quantizations of slices in the affine Grassmannian. Algebra Number Theory **8**(4), 857–893 (2014)
33. I. Losev, Deformations of symplectic singularities and orbit method for semisimple Lie algebras. arXiv:1605.00592
34. J. Lurie, On the classification of topological field theories. Current Developments in Mathematics (Int. Press, Somerville, MA, 2008), pp. 129–280
35. J. Lurie, Derived algebraic geometry VI: $\mathbb{E}[k]$-algebras. http://www.math.harvard.edu/~lurie/papers/DAG-VI.pdf
36. G. Lusztig, Singularities, character formulas, and a q-analogue of weight multiplicities. Astérisque **101–102**, 208–229 (1983)
37. I. Mirković, K. Vilonen, Geometric Langlands duality and representations of algebraic groups over commutative rings. Ann. Math. (2) **166**(1), 95–143 (2007)
38. G.W. Moore, Y. Tachikawa, On 2d TQFTs whose values are holomorphic symplectic varieties, in *Proceedings of Symposia in Pure Mathematics*, vol. 85 (American Mathmatical Society, Providence, 2012), pp. 191–207
39. H. Nakajima, Towards a mathematical definition of Coulomb branches of 3-dimensional $\mathcal{N} = 4$ gauge theories, I. Adv. Theor. Math. Phys. **20**(3), 595–669 (2016)
40. H. Nakajima, Lectures on perverse sheaves on instanton moduli spaces, in *Geometry of Moduli Spaces and Representation Theory*. IAS/Park City Mathematics Series, vol. 24 (American Mathematical Society, Providence, RI, 2017), pp. 381–436
41. H. Nakajima, Questions on provisional Coulomb branches of 3-dimensional $\mathcal{N} = 4$ gauge theories. arXiv:1510.03908

42. H. Nakajima, Introduction to a provisional mathematical definition of Coulomb branches of 3-dimensional $\mathcal{N} = 4$ gauge theories, in *Modern Geometry: A Celebration of the Work of Simon Donaldson*. Proceedings of Symposia in Pure Mathematics, vol. 99 (American Mathematical Society, Providence, RI, 2018), pp. 193–211
43. H. Nakajima, Y. Takayama, Cherkis bow varieties and Coulomb branches of quiver gauge theories of affine type A. Selecta Math. (N.S.) **23**(4), 2553–2633 (2017)
44. S. Raskin, Chiral categories. http://math.mit.edu/~sraskin/chiralcats.pdf
45. S. Raskin, D-modules on infinite-dimensional varieties. http://math.mit.edu/~sraskin/dmod.pdf
46. N. Seiberg, E. Witten, Gauge dynamics and compactification to three dimensions. Advanced Series in Mathematical Physics, vol. 24 (World Scientific, River Edge, NJ, 1997) pp. 333–366
47. B. Webster, Koszul duality between Higgs and Coulomb categories $\mathcal{O}$. arXiv:1611.06541

Chapter 2
Moduli Spaces of Sheaves on Surfaces: Hecke Correspondences and Representation Theory

Andrei Neguţ

Abstract In modern terms, enumerative geometry is the study of **moduli spaces**: instead of counting various geometric objects, one describes the set of such objects, which if lucky enough to enjoy good geometric properties is called a moduli space. For example, the moduli space of linear subspaces of $\mathbb{A}^n$ is the Grassmannian variety, which is a classical object in representation theory. Its cohomology and intersection theory (as well as those of its more complicated cousins, the flag varieties) have long been studied in connection with the Lie algebras $\mathfrak{sl}_n$.

The main point of this mini-course is to make the analogous connection between the moduli space $\mathcal{M}$ of certain more complicated objects, specifically sheaves on a smooth projective surface, with an algebraic structure called the elliptic Hall algebra $\mathcal{E}$ (see Burban and Schiffmann (I Duke Math J 161(7):1171–1231, 2012) and Schiffmann and Vasserot (Duke Math J 162(2):279–366, 2013)). We will recall the definitions of these objects in Sects. 2.1 and 2.2, respectively, but we note that the algebra $\mathcal{E}$ is isomorphic to the quantum toroidal algebra, which is a central-extension and deformation of the Lie algebra $\mathfrak{gl}_1[s^{\pm 1}, t^{\pm 1}]$. Our main result is the following (see Neguţ (*Hecke Correspondences for Smooth Moduli Spaces of Sheaves*. arXiv:1804.03645)):

Theorem 1 *There exists an action $\mathcal{E} \curvearrowright K_{\mathcal{M}}$, defined as in Sect. 2.2.6.*

($K_{\mathcal{M}}$ denotes the algebraic K-theory of the moduli space $\mathcal{M}$, see Sect. 2.1.6) One of the main reasons why one would expect the action $\mathcal{E} \curvearrowright K_{\mathcal{M}}$ is that it generalizes the famous Heisenberg algebra action Grojnowski (Math Res Lett 3(2), 1995) and Nakajima (Ann Math (second series) 145(2):379–388 1997) on the cohomology of Hilbert schemes of points (see Sect. 2.2.1 for a review). In general, such actions are useful beyond the beauty of the structure involved: putting an algebra action on $K_{\mathcal{M}}$ allows one to use representation theory in order to describe various intersection-theoretic computations on $\mathcal{M}$, such as Euler characteristics

A. Neguţ (✉)
Massachusetts Institute of Technology, Department of Mathematics, Cambridge, MA, USA
e-mail: anegut@mit.edu

U. Bruzzo et al. (eds.), *Geometric Representation Theory and Gauge Theory*, Lecture Notes in Mathematics 2248,
https://doi.org/10.1007/978-3-030-26856-5_2

of sheaves. This has far-reaching connections with mathematical physics, where numerous computations in gauge theory and string theory have recently been expressed in terms of the cohomology and K-theory groups of various moduli spaces (the particular case of the moduli space of stable sheaves on a surface leads to the well-known Donaldson invariants). Finally, we will give some hints as to how one would categorify the action of Theorem 1, by replacing the K-theory groups of $\mathcal{M}$ with derived categories of coherent sheaves. As shown in Gorsky et al. (Flag Hilbert schemes, colored projectors and Khovanov-Rozansky homology. arXiv:1608.07308) and Oblomkov and Rozansky (Sel Math New Ser: 1–104), this categorification is closely connected to the Khovanov homology of knots in the 3-sphere or in solid tori, leading one to geometric knot invariants.

2.1 Moduli Spaces of Sheaves on Surfaces

The contents of this section require knowledge of algebraic varieties, sheaves and cohomology, and derived direct and inverse images of morphisms at the level of [10]. Let X be a projective variety over an algebraically closed field of characteristic zero, henceforth denoted by $\mathbb{C}$. We fix an embedding $X \hookrightarrow \mathbb{P}^N$, meaning that the tautological line bundle $\mathcal{O}(1)$ on projective space restricts to a very ample line bundle on X, which we denote by $\mathcal{O}_X(1)$. The purpose of this section is to describe a scheme $\mathcal{M}$ which represents the functor of flat families of coherent sheaves on X, by which we mean the following things:

- for any scheme T, there is an identification:

$$\text{Maps}(T, \mathcal{M}) \cong \left\{ \mathcal{F} \text{ coherent sheaf on } T \times X \text{ which is flat over } T \right\} \tag{2.1}$$

  which is functorial with respect to morphisms of schemes $T \to T'$
- there exists a **universal sheaf** $\mathcal{U}$ on $\mathcal{M} \times X$, by which we mean that the identification in the previous bullet is explicitly given by:

$$T \xrightarrow{\phi} \mathcal{M} \quad \leftrightsquigarrow \quad \mathcal{F} = (\phi \times \text{Id}_X)^*(\mathcal{U}) \tag{2.2}$$

A coherent sheaf $\mathcal{F}$ on $T \times X$ can be thought of as the family of its fibers over closed points $t \in T$, denoted by $\mathcal{F}_t := \mathcal{F}|_{t \times X}$. There are many reasons why one restricts attention to flat families, such as the fact that flatness implies that the numerical invariants of the coherent sheaves $\mathcal{F}_t$ are locally constant in t. We will now introduce the most important such invariant, the Hilbert polynomial.

2.1.1 Subschemes and Hilbert Polynomials

A subscheme of X is the same thing as an ideal sheaf $\mathcal{I} \subset \mathcal{O}_X$, and many classical problems in algebraic geometry involve constructing moduli spaces of subschemes of X with certain properties.

Example 2.1 If $X = \mathbb{P}^N$, the moduli space parametrizing k dimensional linear subspaces of $\mathbb{P}^N$ is the Grassmannian $Gr(k+1, N+1)$.

We will often be interested in classifying subschemes of X with certain properties (in the example above, the relevant properties are dimension and linearity). Many of these properties can be read off algebraically from the ideal sheaf $\mathcal{I}$.

Definition 2.1 The Hilbert polynomial of a coherent sheaf $\mathcal{F}$ on X is defined as:

$$P_{\mathcal{F}}(n) = \dim_{\mathbb{C}} H^0(X, \mathcal{F}(n)) \tag{2.3}$$

for n large enough. We write $\mathcal{F}(n) = \mathcal{F} \otimes \mathcal{O}_X(n)$.

In the setting of the definition above, the Serre vanishing theorem ensures that $H^i(X, \mathcal{F}(n)) = 0$ for $i \geq 1$ and n large enough, which implies that (2.3) is a polynomial in n. A simple exercise shows that if:

$$0 \to \mathcal{F} \to \mathcal{G} \to \mathcal{H} \to 0$$

is a short exact sequence of coherent sheaves on X, then:

$$P_{\mathcal{F}}(n) = P_{\mathcal{G}}(n) - P_{\mathcal{H}}(n)$$

Therefore, fixing the Hilbert polynomial of an ideal sheaf $\mathcal{I} \subset \mathcal{O}_X$ is the same thing as fixing the Hilbert polynomial of the quotient $\mathcal{O}_X/\mathcal{I}$, if X is given.

Example 2.2 If $X = \mathbb{P}^N$ and $\mathcal{I}$ is the ideal sheaf of a k-dimensional linear subspace, then $\mathcal{O}_X/\mathcal{I} \cong \mathcal{O}_{\mathbb{P}^k}$, which implies that:

$$P_{\mathcal{O}_X/\mathcal{I}}(n) = \dim_{\mathbb{C}} H^0(\mathbb{P}^k, \mathcal{O}_{\mathbb{P}^k}(n)) =$$

$$= \dim_{\mathbb{C}} \left\{\text{degree } n \text{ part of } \mathbb{C}[x_0, \ldots, x_k]\right\} = \binom{n+k}{k}$$

If $\mathcal{I}$ is the ideal sheaf of an arbitrary subvariety of $\mathbb{P}^N$, the degree of the Hilbert polynomial $P_{\mathcal{O}_X/\mathcal{I}}$ is the dimension of the subscheme cut out by $\mathcal{I}$, while the leading order coefficient of $P_{\mathcal{O}_X/\mathcal{I}}$ encodes the degree of the said subscheme. Therefore, the Hilbert polynomial knows about geometric properties of subschemes.

2.1.2 Hilbert and Quot Schemes

We have already seen that giving a subscheme of a projective variety X is the same thing as giving a surjective map $\mathcal{O}_X \twoheadrightarrow \mathcal{O}_X/\mathcal{I}$, and that such subschemes are parametrized by their Hilbert polynomials.

Definition 2.2 There exists a moduli space parametrizing subschemes $\mathcal{I} \subset \mathcal{O}_X$ with fixed Hilbert polynomial $P(n)$, and it is called the **Hilbert scheme**:

$$\mathrm{Hilb}_P = \left\{ \mathcal{I} \subset \mathcal{O}_X \text{ such that } P_{\mathcal{O}_X/\mathcal{I}}(n) = P(n) \text{ for } n \gg 0 \right\}$$

We also write:

$$\mathrm{Hilb} = \bigsqcup_{P \text{ polynomial}} \mathrm{Hilb}_P$$

A particularly important case in the setting of our lecture notes is when the Hilbert polynomial $P(n)$ is constant, in which case the subschemes $\mathcal{O}_X/\mathcal{I}$ are finite length sheaves. More specifically, if $P(n) = d$ for some $d \in \mathbb{N}$, then Hilb_P parametrizes subschemes of d points on X. It is elementary to see that Definition 2.2 is the $\mathcal{V} = \mathcal{O}_X$ case of the following more general construction:

Definition 2.3 Fix a coherent sheaf $\mathcal{V}$ on X and a polynomial $P(n)$. There exists a moduli space, called the **Quot scheme**, parametrizing quotients:

$$Quot_{\mathcal{V},P} = \left\{ \mathcal{V} \twoheadrightarrow \mathcal{F} \text{ such that } P_{\mathcal{F}}(n) = P(n) \text{ for } n \gg 0 \right\}$$

We also write:

$$\mathrm{Quot}_{\mathcal{V}} = \bigsqcup_{P \text{ polynomial}} \mathrm{Quot}_{\mathcal{V},P}$$

Definitions 2.2 and 2.3 concern the existence of projective varieties (denoted by Hilb and $\mathrm{Quot}_{\mathcal{V}}$, respectively) which represent the functors of flat families of ideal sheaves $\mathcal{I} \subset \mathcal{O}_X$ and quotients $\mathcal{V} \twoheadrightarrow \mathcal{F}$, respectively. In the language at the beginning of this section, we have natural identifications:

$$\mathrm{Maps}(T, \mathrm{Hilb}) \cong \left\{ \mathcal{I} \subset \mathcal{O}_{T \times X}, \text{ such that } \mathcal{I} \text{ is flat over } T \right\} \tag{2.4}$$

$$\mathrm{Maps}(T, \mathrm{Quot}_{\mathcal{V}}) \cong \left\{ \pi^*(\mathcal{V}) \twoheadrightarrow \mathcal{F}, \text{ such that } \mathcal{F} \text{ is flat over } T \right\} \tag{2.5}$$

where $\mathcal{I}$ and $\mathcal{F}$ are coherent sheaves on $T \times X$, and $\pi : T \times X \to X$ is the standard projection. The flatness hypothesis on these coherent sheaves implies that the Hilbert polynomial of the fibers $\mathcal{O}_X/\mathcal{I}_t$ and $\mathcal{F}_t$ are locally constant functions of the closed point $t \in T$. If these Hilbert polynomials are equal to a given polynomial P, then the corresponding maps in (2.4) and (2.5) land in the connected components $\text{Hilb}_P \subset \text{Hilb}$ and $\text{Quot}_{\mathscr{V},P} \subset \text{Quot}_{\mathscr{V}}$, respectively.

The construction of the schemes Hilb_P and $\text{Quot}_{\mathscr{V},P}$ is explained in Chapter 2 of [11], where the authors also show that (2.2) is satisfied. Explicitly, there exist universal sheaves $\mathscr{I}$ on $\text{Hilb} \times X$ and $\mathscr{F}$ on $\text{Quot}_{\mathscr{V}} \times X$ such that the identifications (2.4) and (2.5) are given by sending a map $\phi : T \to \text{Hilb}, \text{Quot}_{\mathscr{V}}$ to the pull-back of the universal sheaves under ϕ.

Example 2.3 Let us take $X = \mathbb{P}^1$ and consider zero-dimensional subschemes of X. Any such subscheme Z has finite length as an $\mathscr{O}_X$-module, so we may assume this length to be some $d \in \mathbb{N}$. The ideal sheaf of Z is locally principal, hence there exist $[a_1 : b_1], \dots, [a_d : b_d] \in \mathbb{P}^1$ such that $\mathscr{I}$ is generated by:

$$(sa_1 - tb_1) \dots (sa_d - tb_d)$$

where $\mathbb{C}[s, t]$ is the homogeneous coordinate ring of X. Therefore, length d subschemes of $\mathbb{P}^1$ are in one-to-one correspondence with degree d homogeneous polynomials in s, t (up to scalar multiple) and so it should not be a surprise that:

$$\text{Hilb}_d \cong \mathbb{P}^d \tag{2.6}$$

A similar picture holds when X is a smooth curve C, and the isomorphism (2.6) holds locally on the Hilbert scheme of length d subschemes of C.

2.1.3 Moduli Space of Sheaves

If X is a projective variety, the only automorphisms of $\mathscr{O}_X$ are scalars (elements of the ground field $\mathbb{C}$). Because of this, $\text{Hilb} = \text{Quot}_{\mathscr{O}_X}$ is the moduli space of coherent sheaves of the form $\mathscr{F} = \mathscr{O}_X/\mathscr{I}$. Not all coherent sheaves are of this form, e.g. $\mathscr{F} = \mathbb{C}_x \oplus \mathbb{C}_x$ cannot be written as a quotient of $\mathscr{O}_X$ for any closed point $x \in X$. However, Serre's theorem implies that all coherent sheaves $\mathscr{F}$ with fixed Hilbert polynomial can be written as quotients:

$$\phi : \mathscr{O}_X(-n)^{P(n)} \twoheadrightarrow \mathscr{F} \tag{2.7}$$

for some large enough n, where $\mathscr{O}_X(1)$ is the very ample line bundle on X induced from the embedding of $X \hookrightarrow \mathbb{P}^N$ (the existence of (2.7) stems from the fact that $\mathscr{F}(n)$ is generated by global sections, and its vector space of sections has dimension $P(n)$). Therefore, intuitively one expects that the "scheme":

$$\mathscr{M}_P\text{``} := \text{''}\Big\{\mathscr{F} \text{ coherent sheaf on } X \text{ with Hilbert polynomial } P\Big\} \tag{2.8}$$

(in more detail, $\mathcal{M}_P$ should be a scheme with the property that Maps$(T, \mathcal{M}_P)$ is naturally identified with the set of coherent sheaves $\mathcal{F}$ on $T \times X$ which are flat over t, and the Hilbert polynomial of the fibers $\mathcal{F}_t$ is given by P) satisfies:

$$\mathcal{M}_P = \text{Quot}_{\mathcal{O}_X(-n)^{P(n)}, P} / GL_{P(n)} \tag{2.9}$$

where $g \in GL_{P(n)}$ acts on a homomorphism ϕ as in (2.7) by sending it to $\phi \circ g^{-1}$.

The problem with using (2.9) as a definition is that if G is a reductive algebraic group acting on a projective variety Y, it is not always the case that there exists a geometric quotient Y/G (i.e. a scheme whose closed points are in one-to-one correspondence with G-orbits of Y). However, geometric invariant theory ([15]) allows one to define an open subset $Y^{\text{stable}} \subset Y$ of **stable points**, such that Y^{stable}/G is a geometric quotient. The following is proved, for instance, in [11]:

Theorem 2.1 *A closed point (2.7) of $Quot_{\mathcal{O}_X(-n)^{P(n)}, P}$ is stable under the action of $GL_{P(n)}$ from (2.9) if and only if the sheaf $\mathcal{F}$ has the property that:*

$$p_{\mathcal{G}}(n) < p_{\mathcal{F}}(n), \quad \textit{for } n \textit{ large enough}$$

for any proper subsheaf $\mathcal{G} \subset \mathcal{F}$, where the reduced Hilbert polynomial $p_{\mathcal{F}}(n)$ is defined as the Hilbert polynomial $P_{\mathcal{F}}(n)$ divided by its leading order term.

Therefore, putting the previous paragraphs together, there is a scheme:

$$\mathcal{M}_P := \left\{ \mathcal{F} \text{ } \underline{\text{stable}} \text{ coherent sheaf on } X \text{ with Hilbert polynomial } P \right\} \tag{2.10}$$

which is defined as the geometric quotient:

$$\mathcal{M}_P = \text{Quot}^{\text{stable}}_{\mathcal{O}_X(-n)^{P(n)}, P} / GL_{P(n)} \tag{2.11}$$

Moreover, [11] prove that under certain numerical hypotheses (specifically, that the coefficients of the Hilbert polynomial $P(n)$ written in the basis $\binom{n+i-1}{i}$ be coprime integers) there exists a universal sheaf $\mathcal{U}$ on $\mathcal{M}_P \times X$. This sheaf is supposed to ensure that the identification (2.1) is given explicitly by (2.2), and we note that the universal sheaf is only defined up to tensoring with an arbitrary line bundle pulled back from $\mathcal{M}_P$. We fix such a choice throughout this paper.

2.1.4 Tangent Spaces

From now on, let us restrict to the case of moduli spaces $\mathcal{M}$ of stable sheaves over a smooth projective surface S. Then the Hilbert polynomial of any coherent sheaf is completely determined by 3 invariants: the rank r, and the first and second Chern classes c_1 and c_2 of the sheaf. We will therefore write:

$$\mathcal{M}_{(r,c_1,c_2)} \subset \mathcal{M}$$

for the connected component of $\mathcal{M}$ which parametrizes stable sheaves on S with the invariants r, c_1, c_2. For the remainder of this paper, we will make:

$$\textbf{Assumption A}: \quad \gcd(r, c_1 \cdot \mathcal{O}(1)) = 1 \tag{2.12}$$

As explained in the last paragraph of the preceding Subsection, Assumption A implies that there exists a universal sheaf on $\mathcal{M} \times S$.

Exercise 2.1 Compute the Hilbert polynomial of a coherent sheaf $\mathcal{F}$ on a smooth projective surface S in terms of the invariants r, c_1, c_2 of $\mathcal{F}$, the invariants $c_1(S), c_2(S)$ of the tangent bundle of S, and the first Chern class of the line bundle $\mathcal{O}_S(1)$ (Hint: use the Grothendieck-Hirzebruch-Riemann-Roch theorem).

The closed points of the scheme $\mathcal{M}$ are Maps($\mathbb{C}, \mathcal{M}$), which according to (2.1) are in one-to-one correspondence to stable coherent sheaves $\mathcal{F}$ on S. As for the tangent space to $\mathcal{M}$ at such a closed point $\mathcal{F}$, it is given by:

$$\mathrm{Tan}_{\mathcal{F}}\mathcal{M} = \left\{ \text{maps Spec } \frac{\mathbb{C}[\varepsilon]}{\varepsilon^2} \xrightarrow{\Psi} \mathcal{M} \text{ which restrict to Spec } \mathbb{C} \xrightarrow{\mathcal{F}} \mathcal{M} \text{ at } \varepsilon = 0 \right\} \tag{2.13}$$

Under the interpretation (2.1) of Maps($\mathbb{C}[\varepsilon]/\varepsilon^2, \mathcal{M}$), one can prove the following:

Exercise 2.2 The vector space $Tan_{\mathcal{F}}\mathcal{M}$ is naturally identified with $Ext^1(\mathcal{F}, \mathcal{F})$.

It is well-known that a projective scheme (over an algebraically closed field of characteristic zero) is smooth if and only if all of its tangent spaces have the same dimension. Using this, one can prove:

Exercise 2.3 The scheme $\mathcal{M}$ is smooth if the following holds:

$$\textbf{Assumption S}: \quad \begin{cases} \mathcal{K}_S \cong \mathcal{O}_S & \text{or} \\ \mathcal{K}_S \cdot \mathcal{O}(1) < 0 \end{cases} \tag{2.14}$$

Hint: show that the dimension of the tangent spaces $Ext^1(\mathcal{F}, \mathcal{F})$ is locally constant, by using the fact that the Euler correspondence:

$$\chi(\mathcal{F}, \mathcal{F}) = \sum_{i=0}^{2} (-1)^i \dim Ext^i(\mathcal{F}, \mathcal{F})$$

is locally constant, and the fact that stable sheaves $\mathcal{F}$ are simple, i.e. their only automorphisms are scalars (as for $Ext^2(\mathcal{F}, \mathcal{F})$, you may compute its dimension by using Serre duality on a smooth projective surface).

In fact, one can even compute $\chi(\mathscr{F}, \mathscr{F})$ by using the Grothendieck-Hirzebruch-Riemann-Roch theorem. The exact value will not be important to us, but:

Exercise 2.4 Show that (under Assumption S):

$$\dim \mathcal{M}_{(r,c_1,c_2)} = const + 2rc_2 \tag{2.15}$$

where $const$ is an explicit constant that only depends on S, r, c_1 and not on c_2.

2.1.5 *Hecke Correspondences: Part 1*

Fix r and c_1. The moduli space of Hecke correspondences is the locus of pairs:

$$\mathfrak{Z}_1 = \left\{\text{pairs } (\mathscr{F}', \mathscr{F}) \text{ s.t. } \mathscr{F}' \subset \mathscr{F}\right\} \subset \bigsqcup_{c_2 \in \mathbb{Z}} \mathcal{M}_{(r,c_1,c_2+1)} \times \mathcal{M}_{(r,c_1,c_2)} \tag{2.16}$$

In the setting above, the quotient sheaf $\mathscr{F}/\mathscr{F}'$ has length 1, and must therefore be isomorphic to $\mathbb{C}_x$ for some closed point $x \in S$. If this happens, we will use the notation $\mathscr{F}' \subset_x \mathscr{F}$. We conclude that there exist three maps:

$$\begin{array}{ccccc} & & \mathfrak{Z}_1 & & \\ & \swarrow^{p_+} & \downarrow^{p_S} & \searrow^{p_-} & \\ \mathcal{M} & & S & & \mathcal{M} \end{array} \qquad \begin{array}{ccccc} & & (\mathscr{F}' \subset_x \mathscr{F}) & & \\ & \swarrow^{p_+} & \downarrow^{p_S} & \searrow^{p_-} & \\ \mathscr{F}' & & x & & \mathscr{F} \end{array} \tag{2.17}$$

It is not hard to see that the maps p_+, p_-, p_S are all proper. In fact, we have the following explicit fact, which also describes the scheme structure of $\mathfrak{Z}_1$:

Exercise 2.5 The scheme $\mathfrak{Z}_1$ is the projectivization of a universal sheaf:

$$\begin{array}{c} \mathscr{U} \\ \vdots \\ \downarrow \\ \mathcal{M} \times S \end{array} \tag{2.18}$$

in the sense that $\mathbb{P}_{\mathcal{M}\times S}(\mathscr{U}) \cong \mathfrak{Z}_1 \xrightarrow{p_- \times p_S} \mathcal{M} \times S$.

By definition, the projectivization of $\mathscr{U}$ is:

$$\mathbb{P}_{\mathcal{M}\times S}(\mathscr{U}) = \text{Proj}_{\mathcal{M}\times S}\left(\text{Sym}^*(\mathscr{U})\right) \tag{2.19}$$

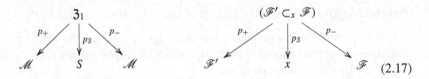

and it comes endowed with a tautological line bundle, denoted by $\mathcal{O}(1)$, and with a map $\rho : \mathbb{P}_{\mathcal{M}\times S}(\mathscr{U}) \to \mathcal{M} \times S$. The scheme (2.19) is completely determined by the fact that maps $\Phi : T \to \mathbb{P}_{\mathscr{M}\times S}(\mathscr{U})$ are in one-to-one correspondence with triples consisting of the following: a map $\phi : T \to \mathscr{M} \times S$, a line bundle $\mathscr{L}$ on T (which will be the pull-back of $\mathscr{O}(1)$ under Φ), and a surjective map $\phi^*(\mathscr{U}) \twoheadrightarrow \mathscr{L}$. However, in the case at hand, we can describe (2.19) a bit more explicitly:

Exercise 2.6 There is a short exact sequence on $\mathscr{M} \times S$:

$$0 \to \mathscr{W} \to \mathscr{V} \to \mathscr{U} \to 0 \tag{2.20}$$

where $\mathscr{V}$ and $\mathscr{W}$ are locally free sheaves on $\mathscr{M} \times S$ (see Proposition 2.2 of [18]).

As a consequence of Exercise 2.6, we have an embedding:

$$\begin{array}{ccc} \mathfrak{Z}_1 \cong \mathbb{P}_{\mathscr{M}\times S}(\mathscr{U}) & \overset{\iota}{\hookrightarrow} & \mathbb{P}_{\mathscr{M}\times S}(\mathscr{V}) \\ & \searrow^{p_- \times p_S} & \downarrow \rho \\ & & \mathscr{M} \times S \end{array} \tag{2.21}$$

which is very helpful, since $\mathbb{P}_{\mathscr{M}\times S}(\mathscr{V})$ is a projective space bundle over $\mathscr{M} \times S$, hence smooth. Moreover, one can even describe the ideal of the embedding ι above. Try to show that it is equal to the image of the map:

$$\rho^*(\mathscr{W}) \otimes \mathscr{O}(-1) \to \rho^*(\mathscr{V}) \otimes \mathscr{O}(-1) \to \mathscr{O} \tag{2.22}$$

on $\mathbb{P}_{\mathscr{M}\times S}(\mathscr{V})$. Therefore, we conclude that $\mathfrak{Z}_1$ is cut out by a section of the vector bundle $\rho^*(\mathscr{W}^\vee) \otimes \mathscr{O}(1)$ on the smooth scheme $\mathbb{P}_{\mathscr{M}\times S}(\mathscr{V})$.

Exercise 2.7 Under Assumption S, $\mathfrak{Z}_1$ is smooth of dimension:

$$const + r(c_2 + c_2') + 1$$

where c_2 and c_2' are the locally constant functions on $\mathfrak{Z}_1 = \{(\mathscr{F}' \subset \mathscr{F})\}$ which keep track of the second Chern classes of the sheaves $\mathscr{F}$ and $\mathscr{F}'$, respectively. The number *const* is the same one that appears in (2.15).

Exercise 2.7 is a well-known fact, which was first discovered for Hilbert schemes more than 20 years ago. You may prove it by describing the tangent spaces to $\mathfrak{Z}_1$ in terms of Ext groups (emulating the isomorphism (2.13) of the previous Subsection), or by looking at Proposition 2.10 of [18]. As a consequence of the dimension estimate in Exercise 2.7, it follows that the section (2.22) is regular, and so $\mathfrak{Z}_1$ is regularly embedded in the smooth variety $\mathbb{P}_{\mathscr{M}\times S}(\mathscr{V})$.

2.1.6 K-Theory and Derived Categories

The schemes $\mathcal{M}$ and $\mathfrak{Z}_1$ will play a major role in what follows, but we must first explain what we wish to do with them. Traditionally, the enumerative geometry of such moduli spaces of sheaves is encoded in their cohomology, but in the present notes we will mostly be concerned with more complicated invariants. First of all, we have their K-theory groups:

$$K_{\mathcal{M}} \quad \text{and} \quad K_{\mathfrak{Z}_1} \tag{2.23}$$

which are defined as the $\mathbb{Q}$-vector spaces generated by isomorphism classes of locally free sheaves on these schemes, modulo the relation $[\mathcal{F}] = [\mathcal{G}] - [\mathcal{H}]$ whenever we have a short exact sequence of locally free sheaves $0 \to \mathcal{F} \to \mathcal{G} \to \mathcal{H} \to 0$.

Example 2.4 K-theory is always a ring, with respect to direct sum and tensor product of vector bundles. In particular, we have a ring isomorphism:

$$K_{\mathbb{P}^n} \to \frac{\mathbb{Q}[\xi]}{(1-\xi)^{n+1}}, \qquad \mathcal{O}(1) \mapsto \xi$$

Functoriality means that if $f : X \to Y$, then there should exist homomorphisms:

$$K_X \underset{f^*}{\overset{f_*}{\rightleftarrows}} K_Y$$

called push-forward and pull-back (or direct image and inverse image, respectively). With our definition, the pull-back f^* is well-defined in complete generality, while the push-forward f_* is well-defined when f is proper and Y is smooth.

Remark 2.1 There is an alternate definition of K-theory called G-theory, where one replaces locally free sheaves by coherent sheaves in the sentence after (2.23). On a smooth projective scheme, the two notions are equivalent because any coherent sheaf has a finite resolution by locally free sheaves. However, G-theory is not in general a ring, but just a module for the K-theory ring. Moreover, G-theory has different functoriality properties from K-theory: the existence of the push-forward f_* only requires $f : X \to Y$ to be a proper morphism (with no restriction on Y), while the pull-back f^* requires f to be an l.c.i. morphism (or at least to satisfy a suitable Tor finiteness condition).

K-theory is a shadow of a more complicated notion, known as the derived category of perfect complexes, which we will denote by:

$$D_{\mathcal{M}} \quad \text{and} \quad D_{\mathfrak{Z}_1}$$

Specifically, the derived category of a projective variety has objects given by complexes of locally free sheaves and morphisms given by maps of complexes, modulo homotopies, and inverting quasi-isomorphisms (in other words, any map of complexes which induces isomorphisms on cohomology is formally considered to be an isomorphism in the derived category). There is a natural map:

$$\text{Obj } D_X \to K_X$$

which sends a complex of locally free sheaves to the alternating sum of its cohomology groups. Since derived categories have, more or less, the same functoriality properties as K-theory groups, we will not review these issues here (but we refer the reader to [12] and [27] for more details, especially the construction of push-forward maps of perfect complexes along l.c.i. morphisms, see Proposition 3.3.20 of [13]). However, we will compare Example 2.4 with the following result, due to Beilinson:

Example 2.5 Any complex in $D_{\mathbb{P}^n}$ is quasi-isomorphic to a complex of direct sums and homological shifts of the line bundles $\{\mathcal{O}, \mathcal{O}(1), \ldots, \mathcal{O}(n)\}$. The Koszul complex:

$$\left[\mathcal{O} \to \mathcal{O}(1)^{\oplus n+1} \to \mathcal{O}(2)^{\oplus \binom{n+1}{2}} \to \ldots \to \mathcal{O}(n)^{\oplus \binom{n+1}{n}} \to \mathcal{O}(n+1)\right]$$

is exact, hence quasi-isomorphic to 0 in $D_{\mathbb{P}^n}$. This is the categorical version of:

$$(1-\xi)^{n+1} = 1-(n+1)\xi+\binom{n+1}{2}\xi^2-\ldots+(-1)^n\binom{n+1}{n}\xi^n+(-1)^{n+1}\xi^{n+1} = 0$$

which is precisely the relation from Example 2.4.

2.2 Representation Theory

2.2.1 *Heisenberg Algebras and Hilbert Schemes*

Consider the Hilbert scheme Hilb_d of d points on a smooth projective algebraic surface S. A basic problem is to compute the Betti numbers:

$$b_i(\text{Hilb}_d) = \dim_{\mathbb{Q}} H^i(\text{Hilb}_d, \mathbb{Q})$$

and their generating function $B(\text{Hilb}_d, t) = \sum_{i \geq 0} t^i b_i(\text{Hilb}_d)$. It turns out that these are easier computed if we consider all d from 0 to ∞ together, as was revealed in the following formula (due to Ellingsrud and Strømme [3] for $S = \mathbb{A}^2$ and then to Göttsche [8] in general):

$$\sum_{d=0}^{\infty} q^d B(\text{Hilb}_d, t) = \prod_{i=1}^{\infty} \frac{(1 + t^{2i-1} q^i)^{b_1(S)} (1 + t^{2i+1} q^i)^{b_3(S)}}{(1 - t^{2i-2} q^i)^{b_0(S)} (1 - t^{2i} q^i)^{b_2(S)} (1 - t^{2i+2} q^i)^{b_4(S)}} \tag{2.24}$$

The reason for the formula above was explained, independently, by Grojnowski [9] and Nakajima [16]. To summarize, they introduced an action of a Heisenberg algebra (to be defined) associated to the surface S on the cohomology group:

$$H = \bigoplus_{d=0}^{\infty} H_d, \qquad \text{where} \qquad H_d = H^*(\text{Hilb}_d, \mathbb{Q}) \tag{2.25}$$

Since the Betti numbers are just the graded dimensions of H, formula (2.24) becomes a simple fact about characters of representations of the Heisenberg algebra. The immediate conclusion is that the representation theory behind the action Heis $\curvearrowright H$ can help one prove numerical properties of H.

Definition 2.4 The **Heisenberg** algebra $Heis$ is generated by infinitely many symbols $\{a_n\}_{n \in \mathbb{Z} \backslash 0}$ modulo the relation:

$$[a_n, a_m] = \delta_{n+m}^0 n \tag{2.26}$$

Let us now describe the way the Heisenberg algebra of Definition 2.4 acts on the cohomology groups (2.25). It is not as straightforward as having a ring homomorphism Heis $\rightarrow$ End(H), but it is morally very close. To this end, let us recall Nakajima's formulation of the action from [16]. Consider the closed subset:

$$\text{Hilb}_{d,d+n} = \Big\{(I, I', x) \text{ such that } I' \subset_x I\Big\} \in \text{Hilb}_d \times \text{Hilb}_{d+n} \times S \tag{2.27}$$

(recall that $I' \subset_x I$ means that the quotient I/I' is a finite length sheaf, specifically length n, supported at the closed point x). We have three natural maps:

$$\begin{array}{ccccc} & & \text{Hilb}_{d,d+n} & & \\ & \overset{p_-}{\swarrow} & \downarrow p_S & \overset{p_+}{\searrow} & \\ \text{Hilb}_d & & S & & \text{Hilb}_{d+n} \end{array} \tag{2.28}$$

While the schemes Hilb_d and S are smooth, $\text{Hilb}_{d,d+n}$ are not for $n > 1$. However, the maps $p_\pm$, p_S are proper, and therefore the following operators are well-defined:

$$H_d \xrightarrow{A_n} H_{d+n} \otimes H_S \qquad\qquad A_n = (p_+ \times p_S)_* \circ p_-^* \tag{2.29}$$

$$H_{d+n} \xrightarrow{A_{-n}} H_d \otimes H_S \qquad A_{-n} = (-1)^{n-1} \cdot (p_- \times p_S)_* \circ p_+^* \tag{2.30}$$

where $H_S = H^*(S, \mathbb{Q})$. We will use the notation $A_{\pm n}$ for the operators above for all d, so one should better think of $A_{\pm n}$ as operators $H \to H \otimes H_S$. Then the main result of Nakajima and Grojnowski states (in a slightly rephrased form):

Theorem 2.2 *We have the following equality of operators $H \to H \otimes H_S \otimes H_S$:*

$$[A_n, A_m] = \delta_{n+m}^0 n \cdot Id_H \otimes [\Delta] \tag{2.31}$$

where in the left hand side we take the difference of the compositions:

$$H \xrightarrow{A_m} H \otimes H_S \xrightarrow{A_n \otimes Id_S} H \otimes H_S \otimes H_S$$

$$H \xrightarrow{A_n} H \otimes H_S \xrightarrow{A_m \otimes Id_S} H \otimes H_S \otimes H_S \xrightarrow{Id_H \otimes swap} H \otimes H_S \otimes H_S$$

and in the right-hand side we multiply by the Poincaré dual class of the diagonal $\Delta \hookrightarrow S \times S$ in $H^*(S \times S, \mathbb{Q}) = H_S \otimes H_S$. The word "swap" refers to the permutation of the two factors of H_S, and the reason it appears is that we want to ensure that in (2.31) *the operators A_n, A_m each act in a single tensor factor of $H_S \otimes H_S$.*

We will refer to the datum $A_n : H \to H \otimes H_S$, $n \in \mathbb{Z} \backslash 0$ as an action of the Heisenberg algebra on H, and relation (2.31) will be a substitute for (2.26). There are two ways one can think about this: the first is that the ring H_S is like the ring of constants for the operators A_n. The second is that one can obtain actual endomorphisms of H associated to any class $\gamma \in H_S$ by the expressions:

$$H \xrightarrow{A_n^\gamma} H \qquad A_n^\gamma = \int_S \gamma \cdot A_n$$

$$H \xrightarrow{A_{-n}^\gamma} H \qquad A_{-n}^\gamma = \int_S \gamma \cdot A_{-n}$$

where $\int_S : H_S \to \mathbb{Q}$ is the integration of cohomology on S. It is not hard to show that (2.31) yields the following commutation relation of operators $H \to H$:

$$[A_n^\gamma, A_m^{\gamma'}] = \delta_{n+m}^0 n \int_S \gamma\gamma' \cdot \mathrm{Id}_H$$

for any classes $\gamma, \gamma' \in H_S$. The relation above is merely a rescaled version of relation (2.26), and it shows that each operator $A_n : H \to H \otimes H_S$ defined by Nakajima and Grojnowski entails the same information as a family of endomorphisms of H indexed by the cohomology group of the surface S itself.

2.2.2 Going Forward

Baranovsky [1] generalized Theorem 2.2 to the setup where Hilbert schemes are replaced by the moduli spaces of stable sheaves:

$$\mathcal{M} = \bigsqcup_{c_2 \in \mathbb{Z}} \mathcal{M}_{(r,c_1,c_2)}$$

from Sect. 2.1.4 (for fixed r, c_1 and with Assumption S in effect). A different generalization entails going from cohomology to K-theory groups, and this is a bit more subtle. The first naive guess is that one should define operators:

$$K_{\mathcal{M}} \xrightarrow{A_n} K_{\mathcal{M}} \otimes K_S$$

for all $n \in \mathbb{Z}\backslash 0$ which satisfy the following natural deformation of relation (2.31):

$$[A_n, A_m] = \delta^0_{n+m} \frac{1-q^{rn}}{1-q} \cdot \mathrm{Id}_{K_{\mathcal{M}}} \otimes [\Delta] \tag{2.32}$$

where q is some invertible parameter (the reason for the appearance of r in the exponent is representation-theoretic, in that K-theory groups of moduli spaces of rank r sheaves yield central charge r representations of Heisenberg algebras). However, we have already said that the "ground ring" should be K_S, so the parameter q should be an invertible element of K_S (it will later turn out that q is the K-theory class of the canonical line bundle of S) and relation (2.32) should read:

$$[A_n, A_m] = \delta^0_{n+m} \mathrm{Id}_{K_{\mathcal{M}}} \otimes \Delta_* \left(\frac{1-q^{rn}}{1-q} \right) \tag{2.33}$$

But even this form of the relation is wrong, mostly because the Künneth formula does not hold (in general) in K-theory: $K_{\mathcal{M} \times S} \not\cong K_{\mathcal{M}} \otimes K_S$. Therefore, the operators we seek should actually be:

$$K_{\mathcal{M}} \xrightarrow{A_n} K_{\mathcal{M} \times S} \tag{2.34}$$

and they must satisfy the following equality of operators $K_{\mathcal{M}} \to K_{\mathcal{M} \times S \times S}$:

$$[A_n, A_m] = \delta^0_{n+m} \Delta_* \left(\frac{1-q^{rn}}{1-q} \cdot \mathrm{proj}^* \right) \tag{2.35}$$

where $\mathrm{proj} : \mathcal{M} \times S \to \mathcal{M}$ is the natural projection. So you may ask whether the analogues of the operators (2.29) and (2.30) in K-theory will do the trick. The answer is no, because the pull-back maps $p_\pm^*$ from (2.28) are not the right objects to study in K-theory. This is a consequence of the fact that the schemes $\mathrm{Hilb}_{d,d+n}$ are very singular for $n \geq 2$, and even if the pull-back maps $p_{\pm}^*$ were defined, then it is not clear what the structure sheaf of $\mathrm{Hilb}_{d,d+n}$ should be replaced with in K-theory, in order to give rise to the desired operators (2.34).

2.2.3 *Framed Sheaves on* $\mathbb{A}^2$

We will now recall the construction of Schiffmann and Vasserot, which generalize the Heisenberg algebra action in the case when $S = \mathbb{A}^2$, in the setting of equivariant K-theory (with respect to the action of the standard torus $\mathbb{C}^* \times \mathbb{C}^* \curvearrowright \mathbb{A}^2$). Historically, this work is based on the computation of K-theoretic Hall algebras by Ginzburg and Vasserot [6], generalized by Varagnolo and Vasserot [28], and then by Nakajima [17] to the general setting of quiver varieties (in the present context, these Hall algebras were studied for the plane in [26] and for the cotangent bundle to a curve in [14]] and [24]).

First of all, since the definition of moduli spaces in the previous section applies to projective surfaces, we must be careful in defining the moduli space $\mathcal{M}$ when $S = \mathbb{A}^2$. The correct definition is the moduli space of framed sheaves on $\mathbb{P}^2$:

$$\mathcal{M} = \left\{ \mathcal{F} \text{ rank } r \text{ torsion-free sheaf on } \mathbb{P}^2, \mathcal{F}|_\infty \overset{\phi}{\cong} \mathcal{O}_\infty^{\oplus r} \right\}$$

where $\infty \subset \mathbb{P}^2$ denotes the divisor at infinity. The space $\mathcal{M}$ is a quasi-projective variety, and we will denote its $\mathbb{C}^* \times \mathbb{C}^*$ equivariant K-theory group by $K_{\mathcal{M}}$. One can define the scheme $\mathfrak{Z}_1$ as in Sect. 2.1.5 and note that it is still smooth. There is a natural line bundle:

$$\begin{array}{c} \mathcal{L} \\ \Big\downarrow \\ \mathfrak{Z}_1 \end{array}$$

whose fiber over a closed point $\{(\mathcal{F}' \subset_x \mathcal{F})\}$ is the one-dimensional space $\mathcal{F}_x / \mathcal{F}'_x$. The maps $p_\pm$ of (2.17) are still well-defined, and they allow us to define operators:

$$K_{\mathcal{M}} \xrightarrow{E_k} K_{\mathcal{M}}, \qquad E_k = p_{+*}\left(\mathcal{L}^{\otimes k} \cdot p_-^*\right) \tag{2.36}$$

for all $k \in \mathbb{Z}$. Note that the map $p_-^* : K_{\mathcal{M}} \to K_{\mathfrak{Z}_1}$ is well-defined in virtue of the smoothness of the spaces $\mathfrak{Z}_1$ and $\mathcal{M}$. The following is the main result of [26]:

Theorem 2.3 *The operators* (2.36) *satisfy the relations in the elliptic Hall algebra.*

To be more precise, [26] define an extra family of operators F_k defined by replacing the signs $+$ and $-$ in (2.36), as well as a family of multiplication operators

H_k, and they show that the three families of operators E_k, F_k, H_k generate the double elliptic Hall algebra. We will henceforth focus only on the algebra generated by the operators $\{E_k\}_{k \in \mathbb{Z}}$ in order to keep things simple, and we will find inside this algebra the positive half of Heis (i.e. the operators (2.34) for $n > 0$).

Remark 2.2 Theorem 2.3 was obtained simultaneously by Feigin and Tsymbaliuk in [5], by using the Ding-Iohara-Miki algebra instead of the elliptic Hall algebra (Schiffmann showed in [25] that the two algebras are isomorphic). We choose to follow the presentation in terms of the elliptic Hall algebra for two important and inter-related reasons: the generators of the elliptic Hall algebra are suitable for categorification and geometry, and the Heisenberg operators (2.34) can be more explicitly described in terms of the elliptic Hall algebra than in terms of the isomorphic Ding-Iohara-Miki algebra.

2.2.4 The Elliptic Hall Algebra

The elliptic Hall algebra $\mathcal{E}$ was defined by Burban and Schiffmann in [2], as a formal model for part of the Hall algebra of the category of coherent sheaves on an elliptic curve over a finite field. The two parameters q_1 and q_2 over which the elliptic Hall algebra are defined play the roles of Frobenius eigenvalues.

Definition 2.5 Write $q = q_1 q_2$. The **elliptic Hall algebra** $\mathcal{E}$ is the $\mathbb{Q}(q_1, q_2)$ algebra generated by symbols $\{a_{n,k}\}_{n \in \mathbb{N}, k \in \mathbb{Z}}$ modulo relations (2.37) and (2.38):

$$[a_{n,k}, a_{n',k'}] = 0 \tag{2.37}$$

if $nk' - n'k = 0$, and:

$$[a_{n,k}, a_{n',k'}] = (1 - q_1)(1 - q_2) \frac{b_{n+n',k+k'}}{1 - q^{-1}} \tag{2.38}$$

if $nk' - n'k = s$ and $\{s, 1, 1\} = \{\gcd(n, k), \gcd(n', k'), \gcd(n + n', k + k')\}$, where:

$$1 + \sum_{s=1}^{\infty} \frac{b_{n_0 s, k_0 s}}{x^s} = \exp\left(\sum_{s=1}^{\infty} \frac{a_{n_0 s, k_0 s}(1 - q^{-s})}{s x^s}\right)$$

for any coprime n_0, k_0.

In full generality, the elliptic Hall algebra defined in [2] has generators $a_{n,k}$ for all $(n, k) \in \mathbb{Z}^2 \backslash (0, 0)$, and relation (2.37) is replaced by a deformed Heisenberg algebra relation between the elements $\{a_{ns,ks}\}_{s \in \mathbb{Z} \backslash 0}$, for any coprime n, k. We refer the reader to Section 2 of [19], where the relations in $\mathcal{E}$ are recalled in our notation. However, we need to know two things about this algebra:

- the operators $\{E_k\}_{k \in \mathbb{Z}}$ of (2.36) will play the role of $a_{1,k}$
- the Heisenberg operators $\{A_n\}_{n \in \mathbb{N}}$ of (2.35) will play the role of $a_{n,0}$

Therefore, the elliptic Hall algebra contains all the geometric operators we have discussed so far, and more. Therefore, the next natural step is to identify the geometric counterparts of the general operators $a_{n,k}$ (which were first discovered in [22] for $S = \mathbb{A}^2$), but the way to do so will require us to introduce the shuffle algebra presentation of the elliptic Hall algebra.

2.2.5 *The Shuffle Algebra*

Shuffle algebras first arose in the context of Lie theory and quantum groups in the work of Feigin and Odesskii [4]. Various instances of this construction have appeared since, and the one we will mostly be concerned with is the following:

Definition 2.6 Let $\zeta(x) = \frac{(1-q_1x)(1-q_2x)}{(1-x)(1-qx)}$. Consider the $\mathbb{Q}(q_1, q_2)$-vector space:

$$\bigoplus_{n=0}^{\infty} \mathbb{Q}(q_1, q_2)(z_1, \dots, z_n)^{Sym} \tag{2.39}$$

endowed with the following shuffle product for any $f(z_1, \dots, z_n)$ and $g(z_1, \dots, z_m)$:

$$f * g = Sym\left[f(z_1, \dots, z_n)g(z_{n+1}, \dots, z_{n+m}) \prod_{n+1 \le j \le n+m}^{1 \le i \le n} \zeta\left(\frac{z_i}{z_j}\right)\right] \tag{2.40}$$

where Sym always refers to symmetrization with respect to all z variables. Then the **shuffle algebra** $\mathscr{S}$ is defined as the $\mathbb{Q}(q_1, q_2)$-subalgebra of (2.39) generated by the elements $\{z_1^k\}_{k \in \mathbb{Z}}$ in the $n = 1$ direct summand.

It was observed in [26] that the map $\mathscr{E} \xrightarrow{\Upsilon} \mathscr{S}$ given by $\Upsilon(a_{1,k}) = z_1^k$ is an isomorphism. The images of the generators $a_{n,k}$ under the isomorphism Υ were worked out in [21], where it was shown that:

$$\Upsilon(a_{n,k}) = \mathrm{Sym}\left[\frac{\prod_{i=1}^n z_i^{\left\lceil \frac{ki}{n} \right\rceil - \left\lceil \frac{k(i-1)}{n} \right\rceil + \delta_i^n - \delta_i^0}}{\left(1 - \frac{qz_2}{z_1}\right) \dots \left(1 - \frac{qz_n}{z_{n-1}}\right)} \prod_{1 \le i < j \le n} \zeta\left(\frac{z_i}{z_j}\right)\right.$$
$$\left.\left(1 + \frac{qz_{a(s-1)+1}}{z_{a(s-1)}} + \frac{q^2 z_{a(s-1)+1} z_{a(s-2)+1}}{z_{a(s-1)} z_{a(s-2)}} + \dots + \frac{q^{s-1} z_{a(s-1)+1} \cdots z_{a+1}}{z_{a(s-1)} \cdots z_a}\right)\right] \tag{2.41}$$

where $s = \gcd(n, k)$ and $a = n/s$. Note that it is not obvious that the elements (2.41) are in the shuffle algebra, and the way [21] proves this fact is by showing that $\mathcal{S}$ coincides with the linear subspace of (2.39) generated by rational functions:

$$\frac{r(z_1, \dots, z_n)}{\prod_{1 \le i \neq j \le n}(z_i - qz_j)}$$

where r goes over all symmetric Laurent polynomials that vanish at $\{z_1, z_2, z_3\} = \{1, q_1, q\}$ and at $\{z_1, z_2, z_3\} = \{1, q_2, q\}$. These vanishing properties are called the **wheel conditions**, following those initially introduced in [4].

Formula (2.41) shows the importance of considering the following elements of $\mathcal{E}$:

$$e_{k_1,\dots,k_n} = \Upsilon^{-1}\left(\text{Sym}\left[\frac{z_1^{k_1}\dots z_n^{k_n}}{\left(1-\frac{qz_2}{z_1}\right)\dots\left(1-\frac{qz_n}{z_{n-1}}\right)}\prod_{1\le i<j\le n}\zeta\left(\frac{z_i}{z_j}\right)\right]\right) \tag{2.42}$$

for any $k_1, \dots, k_n \in \mathbb{Z}$.

Exercise 2.8 Show that the right-hand side of (2.42) lies in $\mathcal{S}$ by showing that it satisfies the wheel conditions that we discussed previously.

Exercise 2.9 Prove the following commutation relations, for all $d, k_1, \dots, k_n \in \mathbb{Z}$:

$$[e_{k_1,\dots,k_n}, e_d] = (1-q_1)(1-q_2)$$
$$\sum_{i=1}^{n}\begin{cases}\sum_{k_i \le a < d} e_{k_1,\dots,k_{i-1},a,k_i+d-a,k_{i+1},\dots,k_n} & \text{if } d > k_i \\ -\sum_{d \le a < k_i} e_{k_1,\dots,k_{i-1},a,k_i+d-a,k_{i+1},\dots,k_n} & \text{if } d < k_i\end{cases} \tag{2.43}$$

There is no summand in the right-hand side corresponding to $k_i = d$. You may prove (2.43) by expressing it as an equality of rational functions in the shuffle algebra $\mathcal{S}$, which you may then prove explicitly (it is not hard, but also not immediate, so try the cases $n \in \{1, 2\}$ first).

It is clear from relations (2.37) and (2.38) that the elements $e_k = a_{1,k}$ generate the algebra $\mathcal{E}$, since any $a_{n,k}$ can be written in terms of sums and products of e_k's. Using the main result of [25], one may show that (2.43) control all relations among the generators $e_k \in \mathcal{E}$. In fact, these relations are over-determined, but we like them because they allow us to express linear combinations of $e_{k_1,\dots,k_n}$ as explicit commutators of e_k's. Therefore, if you have an action of $\mathcal{E}$ where you know how the e_k act, and you wish to prove that the operators $e_{k_1,\dots,k_n}$ act by some formula $(*)$, all you need to do is prove that formula $(*)$ satisfies (2.43).

2.2.6 *The Action of $\mathscr{E}$ on $K_{\mathscr{M}}$*

We will now apply the philosophy in the previous paragraph to the setting of the K-theory group of the moduli space $\mathscr{M}$ of sheaves on a smooth projective surface S (with fixed rank r and first Chern class c_1). As we have seen in Sect. 2.2.2, the way to go is to define operators:

$$K_{\mathscr{M}} \xrightarrow{E_{k_1,\dots,k_n}} K_{\mathscr{M}\times S} \tag{2.44}$$

for all $k_1, \dots, k_n \in \mathbb{Z}$ which satisfy the following analogue of relation (2.43):

$$[E_{k_1,\dots,k_n}, E_d] = \Delta_* \left(\sum_{i=1}^{n} \begin{cases} \sum_{k_i \le a < d} E_{k_1,\dots,k_{i-1},a,k_i+d-a,k_{i+1},\dots,k_n} & \text{if } d > k_i \\ -\sum_{d \le a < k_i} E_{k_1,\dots,k_{i-1},a,k_i+d-a,k_{i+1},\dots,k_n} & \text{if } d < k_i \end{cases} \right) \tag{2.45}$$

as operators $K_{\mathscr{M}} \to K_{\mathscr{M}\times S\times S}$. The left-hand side is defined as in Theorem 2.2, taking care that each of the operators $E_{k_1,\dots,k_n}$ and E_d acts in one and the same factor of $S \times S$. The reason why Δ_* is the natural substitute for $(1-q_1)(1-q_2)$ from (2.43) is that the K-theory class of the diagonal $\Delta \hookrightarrow \mathbb{A}^2 \times \mathbb{A}^2$ is equal to 0 in non-equivariant K-theory, but it is equal to $(1-q_1)(1-q_2)$ equivariantly.

Theorem 2.4 ([20]) *There exist operators* (2.44) *satisfying* (2.45), *with* E_k *given by* (2.36).

In particular, the operators $E_{0,\dots,0}$ all commute with each other, and they will give rise to the K-theoretic version of the positive half of the Heisenberg algebra from Sect. 2.2.1. It is possible to extend Theorem 2.4 to the double of all algebras involved (thus yielding the full Heisenberg) and details can be found in [20].

Given operators $\lambda, \mu : K_{\mathscr{M}} \to K_{\mathscr{M}\times S}$, let us define the following operations:

$$\lambda\mu|_\Delta = \text{composition} \left\{ K_{\mathscr{M}} \xrightarrow{\mu} K_{\mathscr{M}\times S} \xrightarrow{\lambda \boxtimes \mathrm{Id}_S} K_{\mathscr{M}\times S\times S} \xrightarrow{\mathrm{Id}_{\mathscr{M}} \boxtimes \Delta^*} K_{\mathscr{M}\times S} \right\}$$

$$[\lambda, \mu]_{\text{red}} = \nu : K_{\mathscr{M}} \to K_{\mathscr{M}\times S} \quad \text{if } \nu \text{ is such that} \quad [\lambda, \mu] = \Delta_*(\nu)$$

Note that if ν as above exists, it is unique because the map Δ_* is injective (it has a left inverse, i.e. the projection $S \times S \to S$ to one of the factors).

Exercise 2.10 For arbitrary $\lambda, \mu, \nu : K_{\mathscr{M}} \to K_{\mathscr{M}\times S}$, prove the following versions of associativity, the Leibniz rule, and the Jacobi identity, respectively:

$$(\lambda\mu|_\Delta)\nu|_\Delta = \lambda(\mu\nu|_\Delta)|_\Delta$$

$$[\lambda, \mu\nu|_\Delta]_{\text{red}} = [\lambda, \mu]_{\text{red}}\nu|_\Delta + \mu[\lambda, \nu]_{\text{red}}|_\Delta$$

$$[\lambda, [\mu, \nu]_{\text{red}}]_{\text{red}} + [\mu, [\nu, \lambda]_{\text{red}}]_{\text{red}} + [\nu, [\lambda, \mu]_{\text{red}}]_{\text{red}} = 0$$

Theorem 2.4 is an explicit way of saying that the operators (2.44) give rise to an action of the algebra $\mathcal{E}$ on $K_{\mathcal{M}}$ in the sense that there exists a linear map:

$$\Phi : \mathcal{E} \to \mathrm{Hom}(K_{\mathcal{M}}, K_{\mathcal{M}\times S}), \qquad \Phi(e_{k_1,\dots,k_n}) = E_{k_1,\dots,k_n}$$

satisfying the following properties for any $x, y \in \mathcal{E}$:

$$\Phi(xy) = \Phi(x)\Phi(y)|_{\Delta} \tag{2.46}$$

$$\Phi\left(\frac{[x, y]}{(1-q_1)(1-q_2)}\right) = [\Phi(x), \Phi(y)]_{\mathrm{red}} \tag{2.47}$$

The parameters q_1 and q_2 act on K_S as multiplication with the Chern roots of the cotangent bundle Ω^1_S. The reason why the left-hand side of (2.47) makes sense is that for any x, y which are sums of products of the generators $a_{n,k}$ of $\mathcal{E}$, relations (2.37) and (2.38) imply that $[x, y]$ is a multiple of $(1-q_1)(1-q_2)$.

2.3 Proving the Main Theorem

2.3.1 *Hecke Correspondences: Part 2*

As we have seen, the operators E_k of (2.44) are defined by using the correspondence $\mathfrak{Z}_1$ and the line bundle $\mathcal{L}$ on it, as in (2.36). To define the operators $E_{k_1,\dots,k_n}$ in general, we will need to kick up a notch the Hecke correspondences from Sect. 2.1.5, and therefore we will recycle a lot of the notation therein. Thus, $\mathcal{M}$ is still the moduli space of stable sheaves on a smooth projective surface S with fixed r and c_1, satisfying Assumptions A and S from Sect. 2.1.4. Consider:

$$\mathfrak{Z}_2 = \left\{(\mathcal{F}'' \subset \mathcal{F}' \subset \mathcal{F})\right\} \subset \bigsqcup_{c_2\in\mathbb{Z}} \mathcal{M}_{(r,c_1,c_2+2)} \times \mathcal{M}_{(r,c_1,c_2+1)} \times \mathcal{M}_{(r,c_1,c_2)} \tag{2.48}$$

We will denote the support points of flags as above by $x, y \in S$, so that the closed points of $\mathfrak{Z}_2$ take the form $(\mathcal{F}'' \subset_x \mathcal{F}' \subset_y \mathcal{F})$. Consider:

$$\mathfrak{Z}_2 \supset \mathfrak{Z}_2^\bullet = \left\{(\mathcal{F}'' \subset_x \mathcal{F}' \subset_x \mathcal{F}), x \in S\right\}$$

In other words, $\mathfrak{Z}_2^\bullet$ is the closed subscheme of $\mathfrak{Z}_2$ given by the condition that the two support points coincide. These two schemes come endowed with maps:

$$\begin{array}{ccccccc} & \mathfrak{Z}_2 & & & & \mathfrak{Z}_2^\bullet & \\ {}^{\pi_+}\swarrow & & \searrow^{\pi_-} & & {}^{\pi_+^\bullet}\swarrow & & \searrow^{\pi_-^\bullet} \\ \mathfrak{Z}_1 & & \mathfrak{Z}_1 & & \mathfrak{Z}_1 & & \mathfrak{Z}_1 \end{array} \tag{2.49}$$

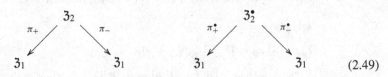

where:

$$\pi_+(\mathscr{F}'' \subset \mathscr{F}' \subset \mathscr{F}) = (\mathscr{F}'' \subset \mathscr{F}'), \qquad \pi_-(\mathscr{F}'' \subset \mathscr{F}' \subset \mathscr{F}) = (\mathscr{F}' \subset \mathscr{F})$$

The maps $\pi_\pm^\bullet$ are given by the same formulas as $\pi_\pm$. The maps $\pi_\pm$ and $\pi_\pm^\bullet$ can be realized as explicit projectivizations, as in (2.19). To see this, let us consider the following coherent sheaves on $\mathfrak{Z}_1 \times S$:

$$\mathscr{W}' = (p_+ \times \mathrm{Id}_S)^*(\mathscr{W}), \; \mathscr{V}' = (p_+ \times \mathrm{Id}_S)^*(\mathscr{V}),$$
$$\mathscr{U}' = (p_+ \times \mathrm{Id}_S)^*(\mathscr{U}) \quad \text{on } \mathfrak{Z}_1 \times S$$
$$\mathscr{W}'^\bullet = (p_+ \times p_S)^*(\mathscr{W}), \; \mathscr{V}'^\bullet = (p_+ \times p_S)^*(\mathscr{V}),$$
$$\mathscr{U}'^\bullet = (p_+ \times p_S)^*(\mathscr{U}) \quad \text{on } \mathfrak{Z}_1$$

More explicitly, we have:

$$\mathscr{U}'_{(\mathscr{F}' \subset_y \mathscr{F}'', x)} = \mathscr{F}'|_x \qquad \mathscr{U}'^\bullet_{(\mathscr{F}' \subset_y \mathscr{F}'')} = \mathscr{F}'|_y$$

and $\mathscr{V}', \mathscr{W}', \mathscr{V}'^\bullet, \mathscr{W}'^\bullet$ are described analogously.

Exercise 2.11 The scheme $\mathfrak{Z}_2$ is the projectivization of $\mathscr{U}'$, in the sense that:

$$\mathbb{P}_{\mathfrak{Z}_1 \times S}(\mathscr{U}') \cong \mathfrak{Z}_2 \xrightarrow{\pi_-} \mathfrak{Z}_1$$

Pull-backs do not preserve short exact sequences in general, because tensor product is not left exact. As a consequence of this phenomenon, it turns out that the short exact sequence (2.20) yields short exact sequences:

$$0 \to \mathscr{W}' \to \mathscr{V}' \to \mathscr{U}' \to 0 \qquad \text{on } \mathfrak{Z}_1 \times S \tag{2.50}$$

$$0 \to \frac{\mathscr{W}'^\bullet}{\mathscr{L}'^\bullet} \to \mathscr{V}'^\bullet \to \mathscr{U}'^\bullet \to 0 \qquad \text{on } \mathfrak{Z}_1 \tag{2.51}$$

where $\mathscr{L}'^\bullet$ is an explicit line bundle that the interested reader can find in Proposition 2.18 of [20]. Therefore, Exercise 2.11 implies that we have diagrams:

$$\begin{array}{ccc} \mathfrak{Z}_2 \cong \mathbb{P}_{\mathfrak{Z}_1}(\mathscr{U}') & \xhookrightarrow{\iota'} & \mathbb{P}_{\mathfrak{Z}_1 \times S}(\mathscr{V}') \\ & \underset{\pi_-}{\searrow} & \downarrow{\scriptstyle \rho'} \\ & & \mathfrak{Z}_1 \times S \end{array} \tag{2.52}$$

$$\begin{array}{ccc} \mathfrak{Z}_2^\bullet \cong \mathbb{P}_{\mathfrak{Z}_1}(\mathscr{U}'^\bullet) & \xhookrightarrow{\iota'^\bullet} & \mathbb{P}_{\mathfrak{Z}_1}(\mathscr{V}'^\bullet) \\ & {\scriptstyle \pi_-^\bullet} \searrow & \downarrow {\scriptstyle \rho'^\bullet} \\ & & \mathfrak{Z}_1 \end{array} \tag{2.53}$$

The ideals of the embeddings ι' and $\iota'^\bullet$ are the images of the maps:

$$\rho'^*(\mathscr{W}') \otimes \mathscr{O}(-1) \to \rho'^*(\mathscr{V}') \otimes \mathscr{O}(-1) \to \mathscr{O} \tag{2.54}$$

$$\rho'^{\bullet *}\left(\frac{\mathscr{W}'^\bullet}{\mathscr{L}'^\bullet}\right) \otimes \mathscr{O}(-1) \to \rho'^{\bullet *}(\mathscr{V}'^\bullet) \otimes \mathscr{O}(-1) \to \mathscr{O} \tag{2.55}$$

on $\mathbb{P}_{\mathfrak{Z}_1 \times S}(\mathscr{V}')$ and $\mathbb{P}_{\mathfrak{Z}_1}(\mathscr{V}'^\bullet)$, respectively. The following Proposition, analogous to Exercise 2.7, implies that the embeddings ι' and $\iota'^\bullet$ are regular. In other words, the compositions (2.54) and (2.55) are duals of regular sections of vector bundles. The regularity of the latter section would fail if we used $\mathscr{W}'^\bullet$ instead of $\mathscr{W}'^\bullet / \mathscr{L}'^\bullet$.

Proposition 2.1 ([20]) *Under Assumption S, $\mathfrak{Z}_2$ and $\mathfrak{Z}_2^\bullet$ have dimensions:*

$$const + r(c_2 + c_2'') + 2 \qquad \text{and} \qquad const + r(c_2 + c_2'') + 1$$

respectively, where c_2 and c_2'' are the locally constant functions on the scheme $\mathfrak{Z}_2 = \{(\mathscr{F}'' \subset \mathscr{F}' \subset \mathscr{F})\}$ which keep track of the second Chern classes of the sheaves $\mathscr{F}$ and $\mathscr{F}''$. Moreover, $\mathfrak{Z}_2$ is an l.c.i. scheme, while $\mathfrak{Z}_2^\bullet$ is smooth.

It is also easy to describe the singular locus of $\mathfrak{Z}_2$: it consists of closed points $(\mathscr{F}'' \subset_x \mathscr{F}' \subset_y \mathscr{F})$ where $x = y$ and the quotient $\mathscr{F}/\mathscr{F}''$ is a split length 2 sheaf (so isomorphic to a direct sum of two skyscraper sheaves $\mathbb{C}_x \oplus \mathbb{C}_x$).

2.3.2 *The Operators*

There are two natural line bundles on the schemes $\mathfrak{Z}_2$ and $\mathfrak{Z}_2^\bullet$, denoted by:

$$\begin{array}{c} \mathscr{L}_1, \mathscr{L}_2 \\ \vdots \\ \downarrow \\ \mathfrak{Z}_2, \mathfrak{Z}_2^\bullet \end{array}$$

whose fiber over a point $\{(\mathscr{F}'' \subset_x \mathscr{F}' \subset_y \mathscr{F})\}$ are the one-dimensional spaces $\mathscr{F}'_x/\mathscr{F}''_x$, $\mathscr{F}_y/\mathscr{F}'_y$, respectively. The maps of (2.17) and (2.49) may be assembled into:

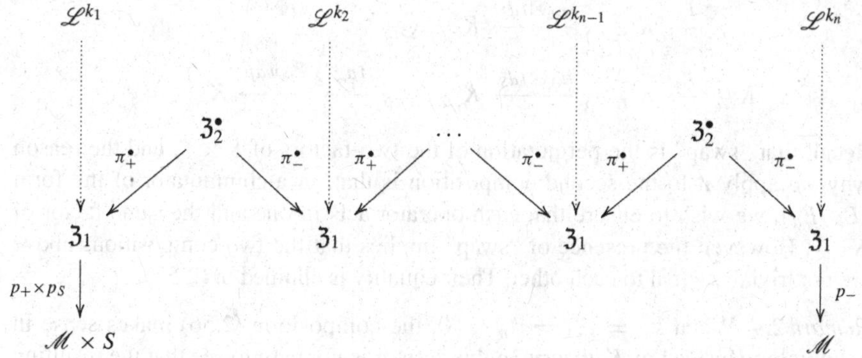

for any $k_1, \dots, k_n \in \mathbb{Z}$. The above diagram of smooth schemes, morphisms and line bundles gives rise to an operator:

$$K_{\mathscr{M}} \xrightarrow{E_{k_1,\dots,k_n}} K_{\mathscr{M}\times S}$$

by tracing pull-back and push-forward maps from bottom right to bottom left, and whenever we reach the scheme $\mathfrak{Z}_1$ for the i–th time, we tensor by the line bundle $\mathscr{L}^{k_{n+1-i}}$. In symbols:

$$E_{k_1,\dots,k_n} = (p_+ \times p_S)_* \left(\mathscr{L}^{k_1} \cdot \pi^\bullet_{+*}\pi^{\bullet *}_-\left(\mathscr{L}^{k_2} \cdot \pi^\bullet_{+*} \dots \pi^{\bullet *}_-\left(\mathscr{L}^{k_{n-1}} \cdot \pi^\bullet_{+*}\pi^{\bullet *}_-\left(\mathscr{L}^{k_n} \cdot p_-^*\right)\dots\right)\right)\right) \tag{2.56}$$

and we claim that these are the operators whose existence was stipulated in Theorem 2.4. Recall that this means that the operators $E_{k_1,\dots,k_n}$ defined as above should satisfy relation (2.45). In the remainder of this lecture, we will prove the said relation in the case $n = 1$, i.e. we will show that:

$$[E_k, E_d] = \Delta_* \left(\begin{cases} \sum_{k \le a < d} E_{a,k+d-a} & \text{if } d > k \\ -\sum_{d \le a < k} E_{a,k+d-a} & \text{if } d < k \end{cases}\right) \tag{2.57}$$

The proof of (2.45) for arbitrary n follows the same principle, although it uses some slightly more complicated geometry and auxiliary spaces.

Remark 2.3 Note that even the case $k = d$ of (2.57), i.e. the relation $[E_k, E_k] = 0$, is non-trivial. The reason for this is that the commutator is defined as the difference of the following two compositions, as in Theorem 2.2:

$$K_{\mathcal{M}} \xrightarrow{E_k} K_{\mathcal{M}\times S} \xrightarrow{E_k \boxtimes Id_S} K_{\mathcal{M}\times S\times S}$$

$$K_{\mathcal{M}} \xrightarrow{E_k} K_{\mathcal{M}\times S} \xrightarrow{E_k \boxtimes Id_S} K_{\mathcal{M}\times S\times S} \xrightarrow{Id_{K_{\mathcal{M}}} \boxtimes swap} K_{\mathcal{M}\times S\times S}$$

Recall that "swap" is the permutation of the two factors of $S \times S$, and the reason why we apply it to the second composition is that, in a commutator of the form $[E_k, E_d]$, we wish to ensure that each operator acts in one and the same factor of $K_{S\times S}$. However, the presence of "swap" implies that the two compositions above are not trivially equal to each other. Their equality is claimed in (2.57).

Remark 2.4 When $k_1 = \ldots = k_n = 0$, the composition (2.56) makes sense in cohomology instead of K-theory. In this case, it is not hard to see that the resulting operator $E_{0,\ldots,0}$ is equal to A_n of (2.29). Indeed, this follows from the fact that the former operator is (morally speaking) given by the correspondence:

$$\left\{(\mathcal{F}' = \mathcal{F}_0 \subset_x \mathcal{F}_1 \subset_x \ldots \subset_x \mathcal{F}_{n-1} \subset_x \mathcal{F}_n = \mathcal{F}, \text{ length } \mathcal{F}_i/\mathcal{F}_{i-1} = 1)\right\} \tag{2.58}$$

between the moduli spaces parametrizing the sheaves $\mathcal{F}$ and $\mathcal{F}'$, while the rank r generalization of the latter operator [1] is given by the correspondence:

$$\left\{(\mathcal{F}' \subset_x \mathcal{F}, \text{ length } \mathcal{F}/\mathcal{F}' = n)\right\} \tag{2.59}$$

Since the correspondence (2.58) is generically 1-to-1 over the correspondence (2.59), this implies that their fundamental classes give rise to the same operators in cohomology. This argument needs care to be made precise, because the operator $E_{0,\ldots,0}$ is not really given by the fundamental class of (2.58), but by some virtual fundamental class that arises from the composition of operators (2.56).

2.3.3 The Moduli Space of Squares

In order to prove (2.57), let us consider the space $\mathcal{Y}$ of quadruples of stable sheaves:

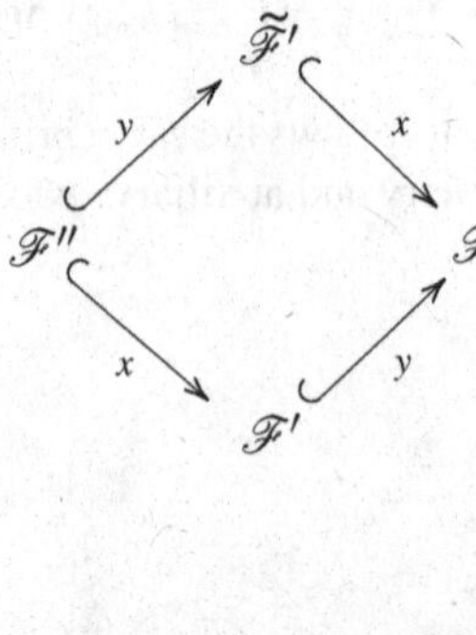

(2.60)

where $x, y \in S$ are arbitrary. There are two maps $\pi^{\downarrow}, \pi^{\uparrow} : \mathscr{Y} \to \mathfrak{Z}_2$ which forget the top-most sheaf and the bottom-most sheaf, respectively, and line bundles:

$$\mathscr{L}_1, \mathscr{L}_2, \mathscr{L}_1', \mathscr{L}_2' \in \mathrm{Pic}(\mathscr{Y})$$

whose fibers are given by the spaces of sections of the length 1 skyscraper sheaves $\mathscr{F}'/\mathscr{F}''$, $\mathscr{F}/\mathscr{F}'$, $\widetilde{\mathscr{F}}'/\mathscr{F}''$, $\mathscr{F}/\widetilde{\mathscr{F}}'$, respectively. Note that:

$$\mathscr{L}_1\mathscr{L}_2 \cong \mathscr{L}_1'\mathscr{L}_2' \tag{2.61}$$

Proposition 2.2 ([20]) *The scheme $\mathscr{Y}$ is smooth, of the same dimension as $\mathfrak{Z}_2$.*

It is easy to see that the map $\mathscr{Y} \xrightarrow{\pi^{\downarrow}} \mathfrak{Z}_2$ is surjective. The fiber of this map above a closed point $(\mathscr{F}'' \subset_x \mathscr{F}' \subset_y \mathscr{F}) \in \mathfrak{Z}_2$ consists of a single point unless $x = y$ and $\mathscr{F}/\mathscr{F}''$ is a split length 2 sheaf, in which case the fiber is a copy of $\mathbb{P}^1$ (**Exercise**: prove this). Since the locus where $x = y$ and $\mathscr{F}/\mathscr{F}''$ is precisely the singular locus of $\mathfrak{Z}_2$, it should not be surprising that $\mathscr{Y}$ is a resolution of singularities of $\mathfrak{Z}_2$. The situation is made even nicer by the following.

Proposition 2.3 ([20]) *We have $\pi^{\downarrow}_*(\mathscr{O}_{\mathscr{Y}}) = \mathscr{O}_{\mathfrak{Z}_2}$ and $R^i\pi^{\downarrow}_*(\mathscr{O}_{\mathscr{Y}}) = 0$ for all $i > 0$.*

The Proposition above can be proved by embedding $\mathscr{Y}$ into $\mathbb{P}_{\mathfrak{Z}_2}(\mathscr{N})$, where $\mathscr{N}$ is the rank 2 vector bundle on $\mathfrak{Z}_2$ with fibers given by $\Gamma(S, \mathscr{F}/\mathscr{F}'')$, and the ideal of this embedding can be explicitly described. As a consequence, one can compute the derived direct images of $\pi^{\downarrow}$ directly.

Exercise 2.12 Find a map of line bundles $\mathscr{L}_1 \xrightarrow{\sigma} \mathscr{L}_2'$ on $\mathscr{Y}$ with zero subscheme:

$$\mathfrak{Z}_2^{\bullet} \cong \{x = y, \mathscr{F}' = \widetilde{\mathscr{F}}'\} \subset \mathscr{Y}$$

2.3.4 *Proof of Relation (2.57)*

Consider the diagram:

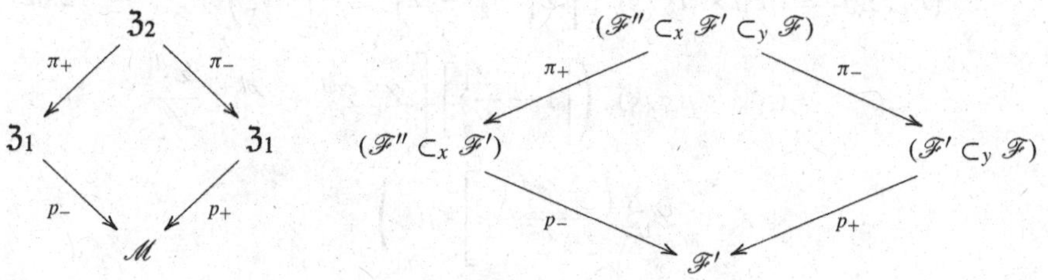

It is clear that the square is Cartesian, but in fact more is true. We know that π_- and p_- are compositions of a regular embedding following by a projective bundle. But comparing (2.21), (2.22) with (2.52), (2.54), we see that the regular embedding and the projective bundle in question are the same for the two maps π_- and p_-, and this implies that the base change formula holds:

$$p_-^* \circ p_{+*} = \pi_{+*} \circ \pi_-^* : K_{\mathfrak{Z}_1} \to K_{\mathfrak{Z}_1}$$

As a consequence of the formula above, one can show the following:

Exercise 2.13 Prove the equality $E_k \circ E_d = (\lambda_+ \times \lambda_{S\times S})_*\left(\mathcal{L}_1^k \mathcal{L}_2^d \cdot \lambda_-^*\right)$, where:

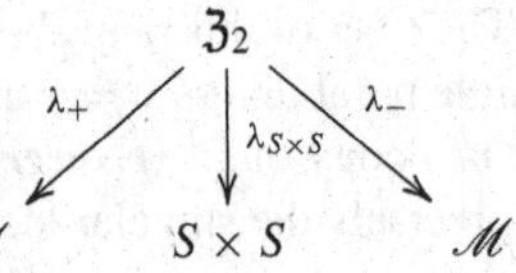

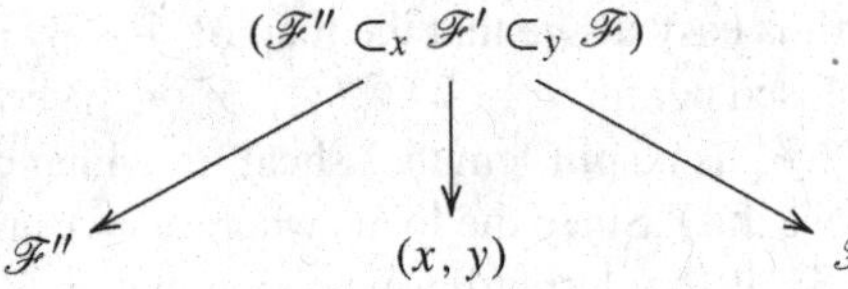

Exercise 2.13 and Proposition 2.3 implies that:

$$E_k \circ E_d = (\mu_+ \times \mu_{S\times S})_*\left(\mathcal{L}_1^k \mathcal{L}_2^d \cdot \mu_-^*\right)$$

$$E_d \circ E_k = (\mu_+ \times \mu_{S\times S})_*\left(\mathcal{L}_1'^d \mathcal{L}_2'^k \cdot \mu_-^*\right)$$

where the maps are as follows:

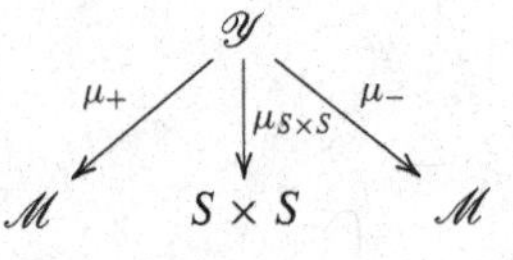

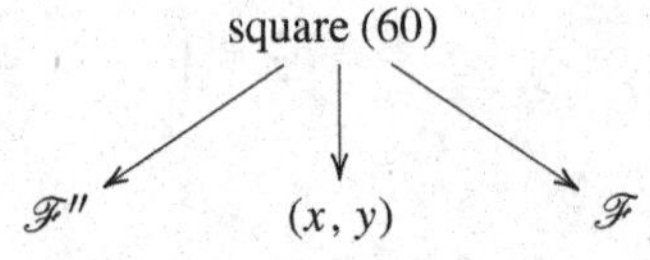

Therefore, assuming $d \geq k$ without loss of generality, we have:

$$[E_k, E_d] = (\mu_+ \times \mu_{S\times S})_*\left(\left[\mathcal{L}_1^k \mathcal{L}_2^d - \mathcal{L}_1'^d \mathcal{L}_2'^k\right] \cdot \mu_-^*\right) \tag{2.62}$$

$$= (\mu_+ \times \mu_{S\times S})_*\left(\left[1 - \frac{\mathcal{L}_1}{\mathcal{L}_2'}\right]\left[\mathcal{L}_1^k \mathcal{L}_2^d + \mathcal{L}_1^{k+1} \frac{\mathcal{L}_2^d}{\mathcal{L}_2'} + \ldots + \mathcal{L}_1^{d-1} \frac{\mathcal{L}_2^d}{\mathcal{L}_2'^{d-k-1}}\right] \cdot \mu_-^*\right)$$

where the last equality is a consequence of (2.61).

Exercise 2.14 Show that Exercise 2.12 implies that:

$$(\mu_+ \times \mu_{S\times S})_* \left(\left[1 - \frac{\mathcal{L}_1}{\mathcal{L}_2'} \right] \cdot \mathcal{L}_1^e \mathcal{L}_2^f \mathcal{L}_1'^g \mathcal{L}_2'^h \cdot \mu_-^* \right)$$

$$= (\nu_+ \times \nu_{S\times S})_* \left(\mathcal{L}_1^{e+g} \mathcal{L}_2^{f+h} \cdot \nu_-^* \right)$$

where the latter maps are as follows:

In terms of the maps (2.17) and (2.49), we have $\nu_\pm = p_\pm \circ \pi_\pm^\bullet|_{\mathfrak{Z}_2^\bullet}$.

Formula (2.62) and Exercise 2.14 imply formula (2.57).

2.3.5 *Toward the Derived Category*

The definition of the operators $E_{k_1,\ldots,k_n}$ in (2.56) immediately generalizes to the derived category (replacing all pull-back and push-forward maps by the corresponding derived inverse and direct image functors), thus yielding functors:

$$D_{\mathcal{M}} \xrightarrow{\widetilde{E}_{k_1,\ldots,k_n}} D_{\mathcal{M}\times S}$$

The proof of the previous Subsection immediately shows how to interpret formula (2.57). Still assuming $d \geq k$, it follows that there exists a natural transformation of functors:

$$\widetilde{E}_d \circ \widetilde{E}_k \to \widetilde{E}_k \circ \widetilde{E}_d$$

whose cone has a filtration with associated graded object:

$$\bigoplus_{a=k}^{d-1} \Delta_* \left(\widetilde{E}_{a,k+d-a} \right)$$

Relation (2.45) has a similar generalization to the derived category, and the proof uses slightly more complicated spaces instead of $\mathcal{Y}$. The corresponding formula leads one to a categorification $\widetilde{\mathcal{E}}$ of the algebra $\mathcal{E}$, which acts on the derived categories of the moduli spaces $\mathcal{M}$. The complete definition of $\widetilde{\mathcal{E}}$ is still work in progress, but when complete, it should provide a categorification of relations (2.37)–(2.38), and in particular a categorification of the Heisenberg algebra (2.26).

Acknowledgements I would like to thank the organizers of the CIME School on Geometric Representation Theory and Gauge Theory: Ugo Bruzzo, Antonella Grassi and Francesco Sala, for making this wonderful event possible. Special thanks are due to Davesh Maulik and Francesco Sala for all their support along the way. I would like to thank the referee for their wonderful suggestions.

References

1. V. Baranovsky, Moduli of sheaves on surfaces and action of the oscillator algebra. J. Differ. Geom. **55**(2), 193–227 (2000)
2. I. Burban, O. Schiffmann, On the hall algebra of an elliptic curve, I. Duke Math. J. **161**(7), 1171–1231 (2012)
3. G. Ellingsrud, S.A. Strømme, On the homology of the Hilbert scheme of points in the plane. Invent. Math. **87**, 343–352 (1987)
4. B. Feigin, A. Odesskii, Quantized moduli spaces of the bundles on the elliptic curve and their applications, in *Integrable Structures of Exactly Solvable Two-Dimensional Models of Quantum Field Theory (Kiev, 2000)*, NATO Science Series, II: Mathematics, Physics and Chemistry, vol. 35 (Kluwer Academic Publishers, Dordrecht, 2001), pp. 123–137
5. B. Feigin, A. Tsymbaliuk, Equivariant K-theory of Hilbert schemes via shuffle algebra. Kyoto J. Math. **51**(4), 831–854 (2011)
6. V. Ginzburg, E. Vasserot, Langlands reciprocity for affine quantum groups of typeA_n. Int. Math. Res. Notices **3**, 67–85 (1993)
7. E. Gorsky, A. Neguţ, J. Rasmussen, Flag Hilbert schemes, colored projectors and Khovanov-Rozansky homology. arXiv:1608.07308
8. L. Göttsche, The Betti numbers of the Hilbert scheme of points on a smooth projective surface. Math. Ann. **286**, 193–207 (1990)
9. I. Grojnowski, Instantons and affine algebras I: The Hilbert scheme and vertex operators. Math. Res. Lett. **3**(2), 1995
10. R. Hartshorne, in *Algebraic Geometry*. Graduate Texts in Mathematics, vol. 52 (Springer, New York, 1977), 978-0-387-90244-9
11. D. Huybrechts, M. Lehn, in *The Geometry of Moduli Spaces of Sheaves*, 2nd edn. (Cambridge University Press, Cambridge 2010). ISBN 978-0-521-13420-0
12. J. Lipman, A. Neeman, Quasi-perfect scheme-maps and boundedness of the twisted inverse image functor. Illinois J. Math. **51**(1), 209–236 (2007)
13. J. Lurie, in *Derived Algebraic Geometry XII: Proper Morphisms, Completions, and the Grothendieck Existence Theorem*. http://www.math.harvard.edu/~lurie/papers/DAG-XII.pdf
14. A. Minets, in *Cohomological Hall Algebras for Higgs Torsion Sheaves, Moduli of Triples and Sheaves on Surfaces*. arXiv:1801.01429
15. D. Mumford, J. Fogarty, F. Kirwan, in *Geometric Invariant Theory*. Ergebnisse der Mathematik und ihrer Grenzgebiete (2) [Results in Mathematics and Related Areas (2)], vo. 34, 3rd edn. (Springer, Berlin, 1994)
16. H. Nakajima, Heisenberg algebra and Hilbert schemes of points on projective surfaces. Ann. Math. (second series) **145**(2), 379–388 (1997)
17. H. Nakajima, Quiver varieties and finite dimensional representations of quantum affine algebras. J. Am. Math. Soc. **14**, 145–238 (2001)
18. A. Neguţ, in *Shuffle Algebras Associated to Surfaces*. arXiv:1703.02027
19. A. Neguţ, in *W-Algebras Associated to Surfaces*. arXiv:1710.03217
20. A. Neguţ, in *Hecke Correspondences for Smooth Moduli Spaces of Sheaves*. arXiv:1804.03645

21. A. Neguţ, The shuffle algebra revisited. Int. Math. Res. Not. **2014**(22), 6242–6275 (2014)
22. A. Neguţ, Moduli of flags of sheaves and their K-theory. Algebr. Geom. **2**(1), 19–43 (2015)
23. A. Oblomkov, L. Rozansky, Knot homology and sheaves on the Hilbert scheme of points on the plane. L. Sel. Math. New Ser. **24**, 2351 (2018)
24. F. Sala, O. Schiffmann, in *Cohomological Hall Algebra of Higgs Sheaves on a Curve*. arXiv:1801.03482
25. O. Schiffmann, Drinfeld realization of the elliptic Hall algebra. J. Algebraic Comb. **35**(2), 237–262 (2012)
26. O. Schiffmann, E. Vasserot, The elliptic Hall algebra and the equivariant K-theory of the Hilbert scheme of $\mathbb{A}^2$. Duke Math. J. **162**(2), 279–366 (2013)
27. B. Toën, in *Proper Local Complete Intersection Morphisms Preserve Perfect Complexes*. arXiv:1210.2827
28. M. Varagnolo, E. Vasserot. On the K-theory of the cyclic quiver variety. Int. Math. Res. Not. **1999**(18), 1005–1028 (1999)

Chapter 3
Notes on Matrix Factorizations and Knot Homology

Alexei Oblomkov

Abstract These are the notes of the lectures delivered by the author at CIME in June 2018. The main purpose of the notes is to provide an overview of the techniques used in the construction of the triply graded link homology. The homology is the space of global sections of a particular sheaf on the Hilbert scheme of points on the plane. Our construction relies on existence on the natural push-forward functor for the equivariant matrix factorizations, we explain the subtleties on the construction in these notes. We also outline a proof of the Markov moves for our homology as well as some explicit localization formulas for knot homology of a large class of links.

3.1 Introduction

The discovery of the knot homology [18] of the links in the three-sphere motivated search for the homological invariants of the three-manifolds. Heegard-Floer homology were discovered soon after Khovanov's seminal work, this homology categorifies the simplest case of WRT invariants (the invariants at the fourth root of unity). More general WRT invariants are beyond of the reach of currently available technique. Thus it is very important to reveal as much structure of the Khovanov homology as it is possible.

The mathematical construction of WRT invariants relies on special properties JW projectors at the root of unity, thus it is natural to search for the analogues of the projectors in the knot homology theory. If the algebraic variety is endowed with the action of the torus with the zero-dimensional locus, the algebraic geometry offers a natural decomposition of the category of coherent sheaves into the mutually orthogonal pieces [13], hence we have a natural analog of the JW projectors. In the paper [25] we constructed a map from the braid group to the category of coherent

A. Oblomkov (✉)
Department of Mathematics and Statistics, University of Massachusetts at Amherst, Amherst, MA, USA
e-mail: oblomkov@math.umass.edu

U. Bruzzo et al. (eds.), *Geometric Representation Theory and Gauge Theory*, Lecture Notes in Mathematics 2248,
https://doi.org/10.1007/978-3-030-26856-5_3

sheaves on the free Hilbert scheme of points on the plane such that Markov moves properties hold for the vector space of the global sections of the sheaf. Thus we have geometric candidate for the JW projectors for such knot homology.

The quest for a geometric interpretation of JW projectors was the main motivation for the author of the notes to develop the connection between sheaves on the Hilbert scheme of points and knot homology. The localization type formulas were first encountered by the author in the joint work with Jake Rasmussen and Vivek Shende [28] where the homology of the torus knots were connected with the topology of the Hilbert schemes of points on the homogeneous plane singularities (see also [10]). However, back in 2012 it was a total mystery to the author how one would expand the relation in [28], [10] beyond the torus knots.

The connection was demystified by Lev Rozansky who was armed with the physics intuition as well as very deep understanding of already existing knot homology theories. As it turned out the searched after knot homology has a natural interpretation within the framework of the Kapustin-Saulina-Rozansky topological quantum field theory for the cotangent bundles to the Lie algebras as targets [22]. A purely mathematical theory underlying the physical predictions is laid out in the series of our joint papers [21, 23–26]. To provide an introduction to the technique of these paper is the main goal of this note.

3.1.1 Main Result

Let us state a consequence of the results from the papers that requires the minimal amount of new notations. We need some notations, though. Throughout the paper we use notation $D^{per}_G(X)$ for the derived category of two-periodic G-equivariant complexes of coherent sheaves on X, where G is a group acting on X. For us particularly important case of the pair X, G is $\mathrm{Hilb}_n(\mathbb{C}^2)$, $T_{sc} = \mathbb{C}^* \times \mathbb{C}^*$ with the scaling action of T_{sc} on $\mathbb{C}^2$. The dual $\mathscr{B}$ to the universal quotient bundle $\mathscr{B}^\vee$, $\mathscr{B}^\vee|_I = \mathbb{C}[x, y]/I$ will be used in our construction of the knot homology.

We also use notation $\mathfrak{Br}_n$ for the braid group on n strands. For an element $\beta \in \mathfrak{Br}_n$ we can form a link in the three-sphere $L(\beta)$ by closing the braid in the most natural way.

Theorem 3.1 ([25]) *There is a constructive procedure that assigns to a braid* $\beta \in \mathfrak{Br}_n$ *an object* $S_\beta \in D^{per}_{T_{sc}}(\mathrm{Hilb}_n(\mathbb{C}^2))$ *such that*

1. $S_{\beta\cdot \mathrm{FT}} = S_\beta \otimes \det(\mathscr{B})$ *where* FT *is the full twist on n strands*
2. *The triply graded vector space* $\mathrm{HHH}(\beta) := H^*(S_\beta \otimes \Lambda^\bullet \mathscr{B})$ *is an isotopy invariant of the closure* $L(\beta)$.
3. *The character of representation of the anti-diagonal torus* $\mathbb{C}^*_a \subset T_{sc}$ *on the spaces* $H^*(S_\beta \otimes \Lambda^i \mathscr{B})$ *is the HOMFLYPT polynomial:*

$$\sum_i a^i \chi_q(\mathbb{C}^*_a, H^*(S_\beta \otimes \Lambda^i \mathscr{B})) = \mathrm{HOMFLYPT}(L(\beta)). \tag{3.1}$$

The constructive procedure in the statement of the theorem relies on the theory of matrix factorizations and in this note we try to present a gentle introduction into the aspects of the theory of matrix factorizations that are necessary for our theory. The author of the notes learned theory of matrix factorizations from discussions with Lev Rozansky, as result the exposition here is quite biased.

The first construction of the triply-graded categorification of the HOMFLYPT invariant appeared in the seminal work of Mikhail Khovanov and Lev Rozansky [19]. It is natural to conjecture that the homology discussed in these notes coincide with the Khovanov-Rozansky homology.

3.1.2 Outline

After defining and motivating the category of matrix factorizations in Sect. 3.3 we spend some time discussing the most common type of matrix factorizations, Koszul matrix factorizations in Sect. 3.2.2. The Koszul matrix factorizations are in many regards are analogous to the complete intersection rings and in this section we make this analogy more precise by providing a method for constructing a matrix factorization from a complete intersection (see Lemma 3.8).

Next we discuss Knorrer periodicity in Sect. 3.2.3 which is the most basic equivalence relation between the categories of matrix factorizations. After that we explain how one would perform push-forward and pull-back between the categories of matrix factorizations, see Sect. 3.2.4. Finally, in the Sect. 3.2.5 we introduce the equivariant matrix factorizations, in particular we explain the difference between the strongly and weakly equivariant matrix factorizations, later we only work with the weakly equivariant matrix factorizations since the weak equivariance allows us to define the equivariant push-forward.

In Sect. 3.3 we explain the key point of our construction, the homomorphism from the braid group $\mathfrak{Br}_n$ to the category of matrix factorizations. First in Sect. 3.3.1 we introduce our main space $\mathscr{X}$ with a potential W and define a convolution algebra structure $\star$ on the category $\mathrm{MF}_{\mathrm{GL}_n\times B^2}(\mathscr{X}, W)$, here $B \subset \mathrm{GL}_n$ is the subgroup of upper-triangular matrices. There is a slightly smaller space $\bar{\mathscr{X}}$ with the potential $\overline{W}$ such that Knorrer periodicity identifies $\mathrm{MF}_{\mathrm{GL}_n\times B^2}(\mathscr{X}, W)$ with $\mathrm{MF}_{B^2}(\bar{\mathscr{X}}, \overline{W})$ and it intertwines the convolution product $\star$ with the convolution product $\bar{\star}$, we provide details in Sect. 3.3.2. After setting notations for the ordinary and affine braid groups in Sect. 3.3.3 we state main properties of the homomorphisms:

$$\Phi : \mathfrak{Br}_n \to \mathrm{MF}_{B^2}(\bar{\mathscr{X}}^{st}, W), \quad \Phi^{aff} : \mathfrak{Br}_n^{aff} \to \mathrm{MF}_{B^2}(\bar{\mathscr{X}}, W),$$

the pull-back along $j_{st} : \mathscr{X}^{st} \to \mathscr{X}$ intertwines these homomorphisms. We postpone the details of the construction of homomorphisms Φ, Φ^{aff} till Sect. 3.5.

In Sect. 3.4 we explain how one can use the homomorphism Φ to construct the triply-graded homology. The free Hilbert scheme FHilb_n^{free} consists of the B-conjugacy classes $\mathrm{FHilb}_n^{free} = \widetilde{\mathrm{FHilb}}_n^{free}/B$ pairs of matrices with a cyclic vector such that the monomials of the matrices applied to the vector span $\mathbb{C}^n$. There is an embedding of the B-cover $\widetilde{\mathrm{FHilb}}_n^{free}$ of the free Hilbert scheme into the stable version of our space $j_e : \widetilde{\mathrm{FHilb}}^{free} \to \bar{\mathscr{X}}^{st}$ and we define the homology group:

$$\mathbb{H}^i(\beta) := \mathbb{H}^*(j_e^*(\Phi(\beta) \otimes \Lambda^i \mathscr{B})^B),$$

where $\mathscr{B}$ is the tautological vector bundle over the free Hilbert scheme. It is shown in [26] that the graded dimension of the total sum

$$\mathrm{HHH}(\beta) = \oplus_i H^i(\beta),$$

is a triply graded knot invariant of the closure $L(\beta)$. We explain in Sect. 3.4.2 why this invariant specializes to the HOMFLYPT invariant after we forget about one of the gradings. Here $H^i(\beta)$ is $\mathbb{H}^{b+i}(\beta)$ with $b = b(\beta)$ being some specific function of β.

The free Hilbert scheme $\mathrm{FHilb}_n^{free} := \widetilde{\mathrm{FHilb}}_n^{free}/B$ is smooth and it contains the usual flag Hilbert scheme $\mathrm{FHilb}_n \subset \mathrm{FHilb}_n^{free}$ which is very singular and not even a local complete intersection. The relation of our homology with the honest flag Hilbert scheme is the following:

$$\mathbb{S}_\beta = j_e(\Phi(\beta))^B \in D^{per}_{T_{sc}}(\mathrm{FHilb}_n^{free}), \quad \mathrm{supp}\left(\mathscr{H}(\mathbb{S}_\beta)\right) \subset \mathrm{FHilb}_n,$$

where $\mathscr{H}(\mathbb{S}_\beta)$ is the sheaf on FHilb_n^{free} which is the homology of the two-periodic complex $\mathbb{S}_\beta$.

The most non-trivial part of the statement from [26] is the fact that the homology $\mathrm{HHH}(\beta)$ does not change under the Markov move that decreases the number of strands in the braid. In Sect. 3.7 we give a sketch of a proof of the Markov move invariance, we rely in this section on the material of Sect. 3.5 where the details of the construction of the braid group action are given.

In the Sect. 3.6 we do a simplest computation in the convolution algebra of the category of matrix factorizations in the case $n = 2$. We show that in $\mathrm{MF}_{B^2}(\bar{\mathscr{X}}^{st}, \overline{W})$ we have an isomorphism

$$\mathscr{C}_\bullet \star \mathscr{C}_\bullet \simeq \mathbf{q}^4\mathscr{C}_\bullet \oplus \mathbf{q}^2\mathscr{C}_\bullet, \tag{3.2}$$

which is the geometric counter-part of the fact that the square of the non-trivial Soergel bimodule for $n = 2$ is equal to the double of itself [33].

Finally, in Sect. 3.8 we define the categorical Chern functor:

$$\mathrm{CH}^{st}_{loc} : \mathrm{MF}_{\mathrm{GL}_n \times B^2}(\mathscr{X}^{st}, W) \to D^{per}_{T_{sc}}(\mathrm{Hilb}_n(\mathbb{C}^2)).$$

We also discuss the properties of the conjugate functor HC^{st}_{loc} (see [27] for the original construction) which is monoidal. The sheaf S_β in the theorem 3.1 is given by:

$$S_\beta = \mathrm{CH}^{st}_{loc}(\Phi(\beta)).$$

The advantage of the sheaf S_β over $\mathbb{S}_\beta$ is that it is a T_{sc}-equivariant periodic complex of sheaves on the smooth manifold $\mathrm{Hilb}_n(\mathbb{C}^2)$ thus we can hope to use T_{sc}-localization technique for computation of the knot homology. There are some technical issues with using the localization method directly as we discuss in Sect. 3.8.5. We also explain how these technical issues could be circumvented and in particular how one can apply this technique to compute the homology of the sufficiently positive elements of Jucy-Murphy algebra. This formula was conjectured in [11].

3.1.3 Other Results

We also would like to mention that many relevant aspects of matrix factorizations are not covered in these notes. The reader could consult papers the original papers of Orlov for the connections with mirror symmetry [29] and paper [5] for some further discussion of the foundations of the theory of matrix factorizations and of course the seminal paper of Khovanov and Rozansky [19] where the first construction of a triply graded homology of the links was proposed. The constructions in these notes are motivated by the physical theory of Kapustin, Saulina, Rozansky [17], the reader is encouraged to read wonderful, basically purely mathematical paper [16] where the role of matrix factorizations in the theory is explained.

Let us also mention that there is a slightly different perspective on the geometric interpretation of the knot homology due to Gorsky, Neguţ, Hogencamp and Rasmussen [9, 11]. Their approach takes the theory of Soergel bimodules and the corresponding link homology construction [19] as a starting point of theory, rather than the categories of matrix factorizations discussed in these notes. Finally, let us mention the recent work of Hogencamp and Elias on categorical diagonalization [6–8] which provides a categorical setting for the localization in the category of coherent sheaves.

These notes by no means were intended as a comprehensive survey of the theory of matrix factorization or of the theory of knot homology. It is a merely is a slightly extended version of the three lectures that the author delivered at 2018 CIME. Thus the author asks for an apology from the colleagues whose contributions to the fields are not covered in the notes.

3.2 Matrix Factorizations

In this section we remind some basic facts about matrix factorizations. There are many excellent expositions on matrix factorizations [4, 5, 29] and we choose not to concentrate on the usual matrix factorizations, instead we aim to define equivariant matrix factorizations and subtleties that arise in an attempt to define such. We also discuss Koszul matrix factorizations and the (equivariant) push-forward functor from [26].

3.2.1 *Motivation and Examples*

Given an affine variety $\mathscr{X}$ and a function F on it we define [4] the homotopy category $\mathrm{MF}(\mathscr{X}, F)$ of matrix factorizations whose objects are complexes of projective $R = \mathbb{C}[\mathscr{X}]$-modules M^0, M^1, $M = M^0 \oplus M^1$ equipped with the differential

$$D = (D^0, D^1) \in \mathrm{Hom}_R(M^0, M^1) \oplus \mathrm{Hom}_R(M^1, M^0)$$

such that $D^2 = F$. Thus $\mathrm{MF}(\mathscr{X}, F)$ is a triangulated category as explained in subsection 3.1 of [29]. We first discuss the objects of this category, then discuss various properties of the morphism spaces.

It is convenient to think about a matrix factorization $(M^0 \oplus M^1, D)$ as a two-periodic curved complex:

$$\dots \xrightarrow{D^1} M^0 \xrightarrow{D^0} M^1 \xrightarrow{D^1} M^0 \xrightarrow{D^0} M^1 \xrightarrow{D^1} \dots, \qquad D^2 = F.$$

Let us look at several basic examples of matrix factorizations and discuss briefly a motivation for the definition of the matrix factorizations by Eisenbud [4].

Example 3.2 $\mathscr{X} = \mathbb{C}$, $R = \mathbb{C}[x]$ and $F = x^5$. The two-periodic complex

$$\dots \xrightarrow{x^2} \underline{R} \xrightarrow{x^3} R \xrightarrow{x^2} R \xrightarrow{x^3} R \xrightarrow{x^2} \dots$$

is an example of an object in $\mathrm{MF}(\mathbb{C}, x^5)$. Here and everywhere below we underline to indicate zeroth homological degree.

Example 3.3 $\mathscr{X} = \mathbb{C}^2$, $R = \mathbb{C}[x, y]$, $F = xy$. The two-periodic complex

$$\dots \xrightarrow{x} \underline{R} \xrightarrow{y} R \xrightarrow{x} R \xrightarrow{y} R \xrightarrow{x} \dots$$

is an example of an object in $\mathrm{MF}(\mathbb{C}^2, xy)$.

The last example has the following geometric interpretation. A module over a quotient ring $Q = \mathbb{C}[x, y]/(xy)$, in general, does not have a finite free resolution. In particular, $M = \mathbb{C}[x] = Q/(y)$ is a module over Q with an infinite free resolution:

$$0 \leftarrow M \overset{y}{\leftarrow} Q \overset{x}{\leftarrow} Q \overset{y}{\leftarrow} Q \overset{x}{\leftarrow} \dots.$$

This resolution has a two-periodic (half-infinite) tail which is a reduction of the matrix factorization from Example 3.3. As explained in [4] this phenomenon is more general.

We felt obliged to mention these results on matrix factorizations to honor the origins of the subjects. For further development of Eisenbud theory the reader is encouraged to look at [4] as well as [29–31] where the connection with the B-model theory is developed. However, the hypersurfaces defined by the potentials from [26] do not have a clear geometric interpretation and it is unclear to us how to make use of Eisenbud's theory in our case. Instead, more elementary homological aspect of the matrix factorizations is important to us. Roughly stated, the very important observation is that all important homological information about the category of matrix factorizations is contained in a neighborhood of the critical locus of the potential. We explain more rigorous statement below.

It is a good place to define morphisms in the category of matrix factorizations. Suppose we have two objects $\mathscr{F}_1 = (M_1, D_1), \mathscr{F}_2 = (M_2, D_2) \in \mathrm{MF}(\mathscr{Z}, F)$ then we define:

$$\underline{\mathrm{Hom}}(\mathscr{F}_1, \mathscr{F}_2) := \{\Psi \in \mathrm{Hom}_R(M_1, M_2) | \Psi \circ D_1 = D_2 \circ \Psi\}.$$

Since the modules M_i are $\mathbb{Z}_2$-graded we have a decomposition

$$\underline{\mathrm{Hom}}(M_1, M_2) = \oplus_{i \in \mathbb{Z}_2} \underline{\mathrm{Hom}}^i(M_1, M_2)$$

where $\underline{\mathrm{Hom}}^i(M_1, M_2) \subset \mathrm{Hom}^i_R(M_1, M_2) := \mathrm{Hom}_R(M_1^0, M_2^i) \oplus \mathrm{Hom}_R(M_1^1, M_2^{i+1})$.

We say that an element $\Psi \in \underline{\mathrm{Hom}}^0(\mathscr{F}_1, \mathscr{F}_2)$ is homotopic to zero: $\Psi \sim 0$ if there is $h \in \mathrm{Hom}^1(M_1, M_2)$ such that $\Psi = h \circ D_1 + D_2 \circ h$. Finally, we define the space of morphisms as a set of equivalence classes with respect to the homotopy equivalence:

$$\mathrm{Hom}(\mathscr{F}_1, \mathscr{F}_2) := \underline{\mathrm{Hom}}^0(\mathscr{F}_1, \mathscr{F}_2)/\sim$$

Now that we defined the objects and morphisms between the objects we can state Orlov's theorem

Theorem 3.4 ([29]) $\mathrm{MF}(\mathscr{Z}, F)$ *has a structure of the triangulated category.*

To complete our discussion of the homological properties of category of matrix factorizations with respect to their critical locus let us observe that an element $f \in R$

naturally gives an element of $\mathrm{Hom}(\mathscr{F}, \mathscr{F})$. For simplicity let us also assume that $\mathscr{X} \subset \mathbb{C}^m$. Then we have a well-defined ideal $I_{crit} \subset R$ generated by $\frac{\partial F}{\partial x_i}$, $i = 1, \dots, m$ and x_i are coordinates on $\mathbb{C}^m$.

Proposition 3.5 *For any $\mathscr{F} \in \mathrm{MF}(\mathscr{X}, F)$ and $f \in I_{crit}$ we have:*

$$\underline{\mathrm{Hom}}^0(\mathscr{F}, \mathscr{F}) \ni f \sim 0.$$

Proof It is enough to show the statement for $f = \frac{\partial F}{\partial x_i}$. Thus the statement follows since:

$$\frac{\partial F}{\partial x_i} = \frac{\partial D}{\partial x_i} D + D \frac{\partial D}{\partial x_i},$$

and $\frac{\partial D}{\partial x_i}$ provides the needed homotopy. □

The last proposition implies that the category of matrix factorizations is model for the coherent sheaves on possibly singular critical locus of the potential F. When the potential is linear in some set of variables then there is an equivalence between with the DG category of the critical locus (see Sect. 3.8.3 for more discussion). Another manifestation of this principle is the shrinking lemma, see Lemma 3.16 below.

3.2.2 *Koszul Matrix Factorizations*

The matrix factorizations from Examples 3.3 and 3.2 are examples of so called *Koszul matrix factorizations* which we discuss in this subsection. Suppose we have a presentation of the potential as sum $F = \sum_{i=1}^n a_i b_i$. Then we define Koszul matrix factorization $\mathrm{K}[\mathbf{a}, \mathbf{b}] \in \mathrm{MF}(\mathscr{X}, F)$ as

$$\mathrm{K}[\mathbf{a}, \mathbf{b}] := (\Lambda^\bullet V, D), \quad D = \sum_i a_i \theta_i + b_i \frac{\partial}{\partial \theta_i},$$

where $V = \langle \theta_1, \dots, \theta_n \rangle$. Examples 3.2, 3.3 are $\mathrm{K}[x^2, x^3]$ and $\mathrm{K}[x, y]$, respectively.

Koszul matrix factorizations are tensor products of the simplest Koszul matrix factorizations. Indeed, given two matrix factorizations $\mathscr{F}_1 \in \mathrm{MF}(\mathscr{X}, F_1)$, $\mathscr{F}_2 \in \mathrm{MF}(\mathscr{X}, F_2)$ the tensor product $\mathscr{F}_1 \otimes \mathscr{F}_2 \in \mathrm{MF}(\mathscr{X}, F_1 + F_2)$ is the matrix factorization $(M_1 \otimes M_2, D_1 \otimes 1 + 1 \otimes D_2)$. Thus we have $\mathrm{K}[\mathbf{a}, \mathbf{b}] = \otimes_{i=1}^n \mathrm{K}[a_i, b_i]$.

An object of the category of matrix factorizations with the zero potential is a two-periodic complex of coherent sheaves. We denote by $D^{per}(\mathscr{X})$ the derived category of the two-periodic complexes of coherent sheaves. Given two matrix factorizations $\mathcal{F}_1 \in \mathrm{MF}(\mathcal{Z}, F)$, $\mathcal{F}_2 \in \mathrm{MF}(\mathcal{Z}, -F)$ their tensor product is an element of $D^{per}(\mathcal{Z})$ and Proposition 3.5 implies:

Corollary 3.6 *For $\mathcal{F}_1 \in \mathrm{MF}(\mathcal{Z}, F)$, $\mathcal{F}_2 \in \mathrm{MF}(\mathcal{Z}, -F)$ homology of the two-periodic complex $\mathcal{F}_1 \otimes \mathcal{F}_2$ are supported on the zero locus of I_{crit}.*

Now let us discuss a method for constructing interesting Koszul matrix factorizations. Let us first recall some basic properties of the usual Koszul complexes. The sequence $f_1, \dots, f_m \in R$ is called *regular* if f_i is not a zero-divisor in the quotient $R/(f_1, \dots, f_{i-1})$ for $i = 1, \dots, n$. It is known that the regularity does not depend on the order of the elements. There is an equivalent way to define regularity with the help of Koszul complexes. The Koszul complex of $\mathbf{f}$ is:

$$\mathrm{K}[\mathbf{f}] = (\Lambda^\bullet V, D), \quad D = \sum_i f_i \frac{\partial}{\partial \theta_i}.$$

Proposition 3.7 *The sequence $(f_1, \dots, f_m)$ is regular if and only if:*

$$H^i(\mathrm{K}[\mathbf{f}]) = 0, \quad i > 0, \qquad H^0(\mathrm{K}(\mathbf{f})) = R/(f_1, \dots, f_m).$$

Given a finite complex of $(C_\bullet, d)$ of free R-modules we denote by $[C_\bullet]_{per}$ the two-periodic folding of the complex. It is an element of $\mathrm{MF}(\mathcal{Z}, 0)$. Suppose $F \in (f_1, \dots, f_m)$ and the sequence $\mathbf{f}$ is regular. Then the lemma below shows that there is an essentially unique way to deform the complex $[\mathrm{K}[\mathbf{f}]]_{per}$ to an element of $\mathrm{MF}(\mathcal{Z}, F)$. We outline a proof of the lemma to demonstrate the key deformation theory technique that is used in many constructions of [26].

Lemma 3.8 *Suppose $F \in (f_1, \dots, f_m)$ and the sequence $\mathbf{f}$ is regular.Then the Koszul complex*

$$C_\bullet = \mathrm{K}[\mathbf{f}] = \{C_0 \xleftarrow{d_1^+} C_1 \xleftarrow{d_2^+} \dots \xleftarrow{d_m^+} C_m\}$$

could be completed with the opposite differentials $d_i^- : C_\bullet \to C_{\bullet+2i-1}$, $i > 0$ such that

$$(C_\bullet, d^+ + d^-) \in \mathrm{MF}(\mathcal{Z}, F).$$

Proof We will construct the differentials d_i iteratively. Since the sequence is regular we have a homotopy equivalence:

$$(C_\bullet, d^+) \sim Q = R/(f_1, \dots, f_m). \tag{3.3}$$

Let us also introduce notation for the graded pieces of the space of homomorphisms:

$$\mathrm{Hom}^i(C_\bullet, C_\bullet) = \oplus_j \mathrm{Hom}(C_j, C_{-i+j}).$$

The element F is an endomorphism of $(C_\bullet, d^+)$ and because of (3.3) it is homotopic to zero by the lemma assumptions. Thus there is a homotopy $h^{(-1)} \in \mathrm{Hom}^{-1}(C_\bullet, C_\bullet)$ such that $F = h^{(1)} \circ d^+ + d^+ \circ h^{(1)}$. Let us set $D^{(1)} = d^+ + h^{(1)}$.

The differential $D^{(1)}$ is the first order approximation for our desired extension. It is not a differential of a matrix factorization if $m > 1$ since:

$$(D^{(1)})^2 = F + (h^{(1)})^2.$$

However the correction term $(h^{(1)})^2$ is actually an element of $\mathrm{Hom}^{-2}_{d^+}(C_\bullet, C_\bullet)$, that is it commutes with the differential d^+:

$$\begin{aligned} d^+ \circ h^{(1)} \circ h^{(1)} &= Fh^{(1)} - h^{(1)} \circ d^+ \circ h^{(1)} = Fh^{(1)} + h^{(1)} \circ h^{(1)} \circ d^+ - h^{(1)} F \\ &= h^{(1)} \circ h^{(1)} \circ d^+. \end{aligned}$$

Thus again by (3.3) there is a homotopy $h^{(3)} \in \mathrm{Hom}^{-3}(C_\bullet, C_\bullet)$ such that $h^{(1)} = d_+ \circ h^{(3)} + h^{(3)} \circ d_+$. We define the next approximation to the needed differential $D^{(3)} = d^+ + h^{(1)} + h^{(3)}$. Again $D^{(3)}$ is not a differential of a matrix factorization if $m > 3$:

$$\begin{aligned} (D^{(1)} + h^{(3)})^2 &= F + (h^{(1)})^2 + D^{(1)} \circ h^{(3)} + h^{(3)} \circ D^{(1)} + (h^{(3)})^2 \\ &= F + h^{(1)} \circ h^{(3)} + h^{(1)} \circ h^{(3)} + (h^{(3)})^2. \end{aligned}$$

The correction term belongs to $\mathrm{Hom}^{<-3}(C_\bullet, C_\bullet)$ and the degree four piece of this term is $h^{(1)} \circ h^{(3)} + h^{(3)} \circ h^{(1)}$. Let us check that $h^{(1)} \circ h^{(3)} + h^{(3)} \circ h^{(1)} \in \mathrm{Hom}^{-4}_{d^+}(C_\bullet, C_\bullet)$:

$$\begin{aligned} d^+ \circ h^{(1)} \circ h^{(3)} &= Fh^{(3)} - h^{(1)} \circ d^+ \circ h^{(3)} = Fh^{(3)} - (h^{(1)})^3 - h^{(1)} \circ h^{(3)} \circ d^+, \\ d^+ \circ h^{(3)} \circ h^{(1)} &= h^{(1)} \circ h^{(1)} \circ h^{(1)} - h^{(3)} \circ d^+ \circ h^{(1)} \\ &= (h^{(1)})^3 - h^{(3)} F - h^{(3)} \circ h^{(1)} \circ d^+. \end{aligned}$$

By the same argument as before homomorphism $h^{(1)} \circ h^{(3)} + h^{(3)} \circ h^{(1)}$ is homotopic to zero and let denote by $h^{(5)} \in \mathrm{Hom}^{-5}(C_\bullet, C_\bullet)$. The next approximation for our differential is $D^{(5)} = d^+ + h^{(1)} + h^{(3)} + h^{(5)}$ and

$$(D^{(5)})^2 - F \in \mathrm{Hom}^{<-5}(C_\bullet, C_\bullet).$$

Similar method could be applied to show that correction term of degree six is homotopic to zero and thus we have the next order correction. Clearly, this iterative procedure terminates since our complex is of finite length. More formal proof of the lemma is given in lemma 2.1 in [26]. □

Remark 3.9 The only assumption on the complex $(C_\bullet, d^+)$ that we used is that

$$\mathrm{Hom}^{<0}_{d^+}(C_\bullet, C_\bullet) \sim 0. \tag{3.4}$$

Thus we can strengthen our lemma a little bit by replacing regularity of the Koszul complex by condition (3.4)

It is natural to ask how canonical is the matrix factorization $(C_\bullet, d^+ + d^-)$ constructed in the previous lemma. Clearly, our method relies on existence of various homotopies which are not unique. However, one can show that the outcome of the iterative procedure in the proof is unique up to an isomorphism. We invite reader to try to apply the iterative method of the previous lemma to show lemma below, a formal proof could be found in lemma 3.7 in [26].

Lemma 3.10 *Let $(C_\bullet, d^+)$ be a complex of free modules with non-trivial terms in degrees from* 0 *to* $l \geq 0$ *such that* $\mathrm{Hom}^{<0}_{d^+}(C_\bullet, C_\bullet) \sim 0$. *Suppose we have two matrix factorizations*

$$\mathscr{F} = (C_\bullet, d^+ + d^-),\ \tilde{\mathscr{F}} = (C_\bullet, d^+ + \tilde{d}^-) \in \mathrm{MF}(\mathscr{Z}, F),$$

where $d^- = \sum_{i\geq 0} d_i^-$, $\tilde{d}^- = \sum_{i\geq 0} \tilde{d}_i^-$, $d_i^-, \tilde{d}_i^- \in \mathrm{Hom}^{-2i-1}(C_\bullet, C_\bullet)$ *and* $F \sim 0$ *as endomorphism of* $(C_\bullet, d^+)$. *Then there is* $\Psi = 1 + \sum_{i>0} \Psi_i$, $\Psi_i \in \mathrm{Hom}^{-i}(C_\bullet, C_\bullet)$ *such that*

$$\Psi \circ (d^+ + d^-) \circ \Psi^{-1} = d^+ + \tilde{d}^-.$$

Because of the previous lemma we will use notation $\mathrm{K}^F(f_1, \ldots, f_m) \in \mathrm{MF}(\mathscr{Z}, F)$ for a matrix factorization from Lemma 3.8.

3.2.3 *Knorrer Periodicity*

The critical locus of the potential $F = xy$ is a point $x = y = 0$ so according to our principle we expect that the category of matrix factorizations with the potential xy is equivalent to the category of matrix factorizations on the point. It is indeed the case and the equivalence is known under the name *Knorrer periodicity* and we explain the details below.

Let us denote the Koszul matrix factorization $\mathrm{K}[x, y] \in \mathrm{MF}(\mathbb{C}^2, xy)$ by K. Then there is an exact functor between triangulated categories:

$$\Phi : \mathrm{MF}(\mathrm{pt}, 0) \to \mathrm{MF}(\mathbb{C}^2, xy), \quad (M, D) \mapsto (M \otimes \mathbb{C}[x, y], D) \otimes \mathrm{K}.$$

The functor in the inverse direction is the restriction functor:

$$\Psi : \mathrm{MF}(\mathbb{C}^2, xy) \to \mathrm{MF}(\mathrm{pt}, 0), \quad (M, D) \mapsto (M|_{x=0}, D|_{x=0}).$$

These functors are mutually inverse. Indeed, to show that $\Psi \circ \Phi = 1$ we observe that $\mathrm{K}|_{x=0} = [\mathbb{C}[y] \xleftarrow{y} \mathbb{C}[y]]$ which is a sky-scarper at $y = 0$. We leave it as an exercise to a reader to show $\Phi \circ \Psi = 1$.

More generally, if $\mathscr{Z} = \mathscr{Z}_0 \times \mathbb{C}^2_{x,y}$ and $F_0 \in \mathbb{C}[\mathscr{Z}_0]$ then there is a functor: $\Phi : \mathrm{MF}(\mathscr{Z}_0, F_0) \to \mathrm{MF}(\mathscr{Z}, F_0 + xy)$ given by tensoring with the Koszul complex $\mathrm{K}[x, y]$.

Theorem 3.11 ([29]) *The functor Φ is an equivalence of triangulated categories.*

3.2.4 *Functoriality*

Now we will use previously developed technique to define the push-forward functor for matrix factorizations. First we discuss a construction of the push-forward for an embedding map $j : \mathscr{Z}_0 \hookrightarrow \mathscr{Z}$ where $\mathscr{Z} = Spec(S)$ and $\mathscr{Z}_0 = Spec(R)$, $R = S/I$.

Theorem 3.12 ([26]) *Suppose we have $F \in S$, $F_0 = j^*(F)$ and $I = (f_1, \dots, f_m)$ where f_i form a regular sequence. Then there is well-defined functor of triangulated categories:*

$$j_* : \mathrm{MF}(\mathscr{Z}_0, F_0) \to \mathrm{MF}(\mathscr{Z}, F)$$

Given an element $\mathscr{F} = (M, D) \in \mathrm{MF}(\mathscr{Z}_0, F_0)$ let us explain the construction of the element $j_*(\mathscr{F}) = \widetilde{F} \in \mathrm{MF}(\mathscr{Z}, F)$. Since $M = R^n$ for some n we can lift it to the module $\widetilde{M} = S^n$ as well the differential to a $\mathbb{Z}_2$-graded endomorphism $\widetilde{D} \in \mathrm{Hom}_S(S^n, S^n)$, $\widetilde{D}|_{\mathscr{Z}_0} = D$. Since f_i form a regular sequence we can form Koszul complex $\mathrm{K}(f_1, \dots, f_m) = (\Lambda^\bullet \mathbb{C}^n \otimes S, d_K)$ which is a resolution of S-module R. The technique similar to the method of Lemma 3.8 yields

Lemma 3.13 ([26]) *There are d^-_{ij}: $\tilde{M} \otimes \Lambda^i \mathbb{C}^n \otimes S \to \tilde{M} \otimes \Lambda^j \mathbb{C}^n \otimes S$, $i - j \in \mathbb{Z}_{>0}$ such that*

$$\widetilde{\mathscr{F}} = (\Lambda^\bullet \mathbb{C}^n \otimes S, d_K + \widetilde{D} + d^-) \in \mathrm{MF}(\mathscr{Z}, F)$$

and the element $\widetilde{\mathscr{F}}$ is unique up to isomorphism.

To complete proof of the Theorem 3.12 we need to show that the construction of j^* extends to the spaces of the morphisms between the objects and to the space of homotopies between the morphism, it is shown in lemma 3.7 of [26] and we refer interested reader there for the technical details.

Unlike push-forward the pull-back functor is rather elementary. Suppose we have $f : \mathscr{Z} \to \mathscr{Z}_0$ a morphism of affine varieties and $F = f^*(F_0)$, $F_0 \in \mathbb{C}[\mathscr{Z}_0]$. Since pull-back of a free module is free, we have a well-defined functor:

$$f^* : \mathrm{MF}(\mathscr{Z}_0, F_0) \to \mathrm{MF}(\mathscr{Z}, F).$$

Finally, let us remark that the above defined pull-back and push-forward functors satisfy the smooth base change relation for commuting squares of maps.

3.2.5 *Equivariant Matrix Factorizations*

A matrix factorization is a natural object attached to a function on the affine manifolds. However limiting yourself to only affine manifolds is frustrating, so one would like to develop a theory of matrix factorizations on quasi-projective manifolds. There are some proposals in the literature for such theory, see for example [32].

In our work [26] we chose an approach that is probably more limited than the one from [32] but has an advantage of being computation friendly. So in [26] to explore matrix factorizations on the manifolds that are group quotients of the affine manifolds, we develop theory of equivariant matrix factorizations. In this section we motivate our definitions and outline the ingredients of the construction from [26].

Suppose the affine manifold $\mathscr{X}$ has an action of an algebraic group H and $F \in \mathbb{C}[\mathscr{X}]^H$. Then one can say the matrix factorization $\mathscr{F} = (M, D) \in \mathrm{MF}(\mathscr{X}, F)$ is *strongly H-equivariant* if M is endowed with H-representation structure and the differential D is H-equivariant. Let us denote the set of strongly equivariant matrix factorizations by $\mathrm{MF}_H^{str}(\mathscr{X}, F)$. By requiring the morphism between the objects and the homotopies between the morphisms to be H-equivariant we can provide $\mathrm{MF}_H^{str}(\mathscr{X}, F)$ with the structure of the triangulated category.

However, the notion of strong equivariance turns out to be too restrictive. Indeed, one of the key tools in our arsenal is the extension Lemma 3.8 together with the push-forward functor. So we would like to have an analog of Lemma 3.8 in the equivariant setting, for the H-equivariant ideal $I = (f_1, \dots, f_m)$ with f_i forming a regular sequence. Unfortunately, the proof of the lemma fails in the strongly equivariant setting because we can not guarantee that the homotopies in the iterative construction of proof are equivariant. If H is reductive, we can save the proof by averaging along the maximal compact subgroup of H. But for a non-reductive group we need a weaker notion of equivariance that relies on the Chevalley-Eilenberg complex explained below.

Let $\mathfrak{h}$ be the Lie algebra of H. Chevalley-Eilenberg complex $\mathrm{CE}_{\mathfrak{h}}$ is the complex $(V_\bullet(\mathfrak{h}), d)$ with $V_p(\mathfrak{h}) = U(\mathfrak{h}) \otimes_{\mathbb{C}} \Lambda^p \mathfrak{h}$ and differential $d_{ce} = d_1 + d_2$ where:

$$d_1(u \otimes x_1 \wedge \cdots \wedge x_p) = \sum_{i=1}^{p} (-1)^{i+1} u x_i \otimes x_1 \wedge \cdots \wedge \hat{x}_i \wedge \cdots \wedge x_p,$$

$$d_2(u \otimes x_1 \wedge \cdots \wedge x_p) = \sum_{i<j} (-1)^{i+j} u \otimes [x_i, x_j] \wedge x_1 \wedge \cdots \wedge \hat{x}_i \wedge \cdots \wedge \hat{x}_j \wedge \cdots \wedge x_p,$$

Let us denote by Δ the standard map $\mathfrak{h} \to \mathfrak{h} \otimes \mathfrak{h}$ defined by $x \mapsto x \otimes 1 + 1 \otimes x$. Suppose V and W are modules over the Lie algebra $\mathfrak{h}$ then we use notation

$V\overset{\Delta}{\otimes}W$ for the $\mathfrak{h}$-module which is isomorphic to $V \otimes W$ as a vector space, the $\mathfrak{h}$-module structure being defined by Δ. Respectively, for a given $\mathfrak{h}$-equivariant matrix factorization $\mathcal{F} = (M, D)$ we denote by $\mathrm{CE}_{\mathfrak{h}}\overset{\Delta}{\otimes}\mathcal{F}$ the $\mathfrak{h}$-equivariant matrix factorization $(CE_{\mathfrak{h}}\overset{\Delta}{\otimes}\mathcal{F}, D + d_{ce})$. The $\mathfrak{h}$-equivariant structure on $\mathrm{CE}_{\mathfrak{h}}\overset{\Delta}{\otimes}\mathcal{F}$ originates from the left action of $U(\mathfrak{h})$ that commutes with right action on $U(\mathfrak{h})$ used in the construction of $\mathrm{CE}_{\mathfrak{h}}$.

A slight modification of the standard fact that $\mathrm{CE}_{\mathfrak{h}}$ is the resolution of the trivial module implies that $\mathrm{CE}_{\mathfrak{h}}\overset{\Delta}{\otimes}M$ is a free resolution of the $\mathfrak{h}$-module M.

Now we about to define a new category whose objects we refer to as *weakly equivariant matrix factorizations*. The objects of this category $\mathrm{MF}_{\mathfrak{h}}(\mathcal{Z}, W)$ are triples:

$$\mathcal{F} = (M, D, \partial), \quad (M, D) \in \mathrm{MF}(\mathcal{Z}, W)$$

where $M = M^0 \oplus M^1$ and $M^i = \mathbb{C}[\mathcal{Z}] \otimes V^i$, $V^i \in \mathrm{Mod}_{\mathfrak{h}}$, $\partial \in \oplus_{i>j}\mathrm{Hom}_{\mathbb{C}[\mathcal{Z}]}(\Lambda^i\mathfrak{h} \otimes M, \Lambda^j\mathfrak{h} \otimes M)$ and D is an odd endomorphism $D \in \mathrm{Hom}_{\mathbb{C}[\mathcal{Z}]}(M, M)$ such that

$$D^2 = F, \quad D_{tot}^2 = F, \quad D_{tot} = D + d_{ce} + \partial,$$

where the total differential D_{tot} is an endomorphism of $\mathrm{CE}_{\mathfrak{h}}\overset{\Delta}{\otimes}M$, that commutes with the $U(\mathfrak{h})$-action.

Note that we do not impose the equivariance condition on the differential D in our definition of matrix factorizations. On the other hand, if $\mathcal{F} = (M, D) \in \mathrm{MF}^{str}(\mathcal{Z}, F)$ is a matrix factorization with D that commutes with $\mathfrak{h}$-action on M then $(M, D, 0) \in \mathrm{MF}_{\mathfrak{h}}(\mathcal{Z}, F)$.

There is a forgetful map for the objects of the categories $\mathrm{Ob}(\mathrm{MF}_{\mathfrak{h}}(\mathcal{Z}, F)) \to \mathrm{Ob}(\mathrm{MF}(\mathcal{Z}, F))$ that forgets about the correction differentials:

$$\mathcal{F} = (M, D, \partial) \mapsto \mathcal{F}^{\sharp} := (M, D).$$

Given two $\mathfrak{h}$-equivariant matrix factorizations $\mathcal{F} = (M, D, \partial)$ and $\tilde{\mathcal{F}} = (\tilde{M}, \tilde{D}, \tilde{\partial})$ the space of morphisms $\mathrm{Hom}(\mathcal{F}, \tilde{\mathcal{F}})$ consists of homotopy equivalence classes of elements $\Psi \in \mathrm{Hom}_{\mathbb{C}[\mathcal{Z}]}(\mathrm{CE}_{\mathfrak{h}}\overset{\Delta}{\otimes}M, \mathrm{CE}_{\mathfrak{h}}\overset{\Delta}{\otimes}\tilde{M})$ such that $\Psi \circ D_{tot} = \tilde{D}_{tot} \circ \Psi$ and Ψ commutes with $U(\mathfrak{h})$-action on $\mathrm{CE}_{\mathfrak{h}}\overset{\Delta}{\otimes}M$. Two maps $\Psi, \Psi' \in \mathrm{Hom}(\mathcal{F}, \tilde{\mathcal{F}})$ are homotopy equivalent if there is

$$h \in \mathrm{Hom}_{\mathbb{C}[\mathcal{Z}]}(\mathrm{CE}_{\mathfrak{h}}\overset{\Delta}{\otimes}M, \mathrm{CE}_{\mathfrak{h}}\overset{\Delta}{\otimes}\tilde{M})$$

such that $\Psi - \Psi' = \tilde{D}_{tot} \circ h - h \circ D_{tot}$ and h commutes with $U(\mathfrak{h})$-action on $\mathrm{CE}_{\mathfrak{h}}\overset{\Delta}{\otimes}M$.

Given two $\mathfrak{h}$-equivariant matrix factorizations $\mathcal{F} = (M, D, \partial) \in \mathrm{MF}_{\mathfrak{h}}(\mathcal{Z}, F)$ and $\tilde{\mathcal{F}} = (\tilde{M}, \tilde{D}, \tilde{\partial}) \in \mathrm{MF}_{\mathfrak{h}}(\mathcal{Z}, \tilde{F})$ we define $\mathcal{F} \otimes \tilde{\mathcal{F}} \in \mathrm{MF}_{\mathfrak{h}}(\mathcal{Z}, F + \tilde{F})$ as an equivariant matrix factorization $(M \otimes \tilde{M}, D + \tilde{D}, \partial + \tilde{\partial})$.

We define an embedding-related push-forward in the case when the subvariety $\mathcal{Z}_0 \overset{j}{\hookrightarrow} \mathcal{Z}$ is the common zero of an ideal $I = (f_1, \ldots, f_n)$ such that the functions $f_i \in \mathbb{C}[\mathcal{Z}]$ form a regular sequence. We assume that the Lie algebra $\mathfrak{h}$ acts on $\mathcal{Z}$ and I is $\mathfrak{h}$-invariant. In section 3 of [26] we use technique similar to the proof of lemma 3.8 to show that there is a well-defined functor:

$$j_*\colon \mathrm{MF}_{\mathfrak{h}}(\mathcal{Z}_0, W|_{\mathcal{Z}_0}) \longrightarrow \mathrm{MF}_{\mathfrak{h}}(\mathcal{Z}, W),$$

for any $\mathfrak{h}$-invariant element $W \in \mathbb{C}[\mathcal{Z}]^{\mathfrak{h}}$.

For our construction of the convolution algebras we also need to define the equivariant push-forward along a projection. Suppose $\mathcal{Z} = \mathcal{X} \times \mathcal{Y}$, both $\mathcal{Z}$ and $\mathcal{X}$ have $\mathfrak{h}$-action and the projection $\pi : \mathcal{Z} \to \mathcal{X}$ is $\mathfrak{h}$-equivariant. Then for any $\mathfrak{h}$ invariant element $w \in \mathbb{C}[\mathcal{X}]^{\mathfrak{h}}$ there is a functor $\pi_*\colon \mathrm{MF}_{\mathfrak{h}}(\mathcal{Z}, \pi^*(w)) \to \mathrm{MF}_{\mathfrak{h}}(\mathcal{X}, w)$ which simply forgets the action of $\mathbb{C}[\mathcal{Y}]$.

Finally, let us discuss the quotient map. The complex $\mathrm{CE}_{\mathfrak{h}}$ is a resolution of the trivial $\mathfrak{h}$-module by free modules. Thus the correct derived version of taking $\mathfrak{h}$-invariant part of the matrix factorization $\mathcal{F} = (M, D, \partial) \in \mathrm{MF}_{\mathfrak{h}}(\mathcal{Z}, W)$, $W \in \mathbb{C}[\mathcal{Z}]^{\mathfrak{h}}$ is

$$\mathrm{CE}_{\mathfrak{h}}(\mathcal{F}) := (\mathrm{CE}_{\mathfrak{h}}(M), D + d_{ce} + \partial) \in \mathrm{MF}(\mathcal{Z}/H, W),$$

where $\mathcal{Z}/H := \mathrm{Spec}(\mathbb{C}[\mathcal{Z}]^{\mathfrak{h}})$ and use the general definition for an $\mathfrak{h}$-module V:

$$\mathrm{CE}_{\mathfrak{h}}(V) := \mathrm{Hom}_{\mathfrak{h}}(\mathrm{CE}_{\mathfrak{h}}, \mathrm{CE}_{\mathfrak{h}} \overset{\Delta}{\otimes} V).$$

3.3 Braid Groups and Matrix Factorizations

In this section we explain a construction for an action of the finite and affine braid groups on the particular categories of the matrix factorizations from [26]. First we explain a construction for the convolution algebra on our categories of matrix factorizations. Then we explain a categorification of the homomorphism from the affine braid group to the finite braid group from [24].

3.3.1 Convolution Product

Let us first motivate the definition of the space that host our categories of matrix factorizations. Somewhat abusing notations we introduce space $\sqrt{\mathcal{X}} = \mathfrak{gl}_n \times \mathrm{GL}_n \times$

$\mathfrak{n}$ where $\mathfrak{n}$ stands for the Lie algebra of strictly upper-triangular matrices. We omit the sub-index since the size of the matrices is clear from the context, we also use G and $\mathfrak{g}$ for GL_n and $\mathfrak{gl}_n$ in this situation.

The space $\sqrt{\mathscr{X}}$ has the action of the group of upper-triangular matrices B and G:

$$(h, b) \cdot (X, g, Y) = (\mathrm{Ad}_h(X), hgb, \mathrm{Ad}_b^{-1} Y), \quad (h, b) \in G \times B.$$

The flag variety Fl is a quotient G/B since every full flag can be moved into the standard flag by G-action and B is the stabilizer group of the standard flag. The group B acts on the tangent space to Fl at the point of standard flag and as B-module the tangent space is equal $\mathfrak{n}$. Thus the B-quotient of $\sqrt{\mathscr{X}}$ is naturally isomorphic to the cotangent bundle to the flag variety:

$$\sqrt{\mathscr{X}}/B = \mathfrak{g} \times T^*\mathrm{Fl}$$

Thus G-action on $\sqrt{\mathscr{X}}$ induces the G-action on $\mathfrak{g} \times T^*\mathrm{Fl}$.

The space $T^*\mathrm{Fl}$ is symplectic and the G-action preserves the symplectic form. Thus there is a moment map $\mu : T^*\mathrm{Fl} \to \mathfrak{g}^*$. The trace identifies $\mathfrak{g}$ with $\mathfrak{g}^*$ and we can think of the moment map as a $\mathfrak{g}$-linear B-invariant function:

$$\mu : \sqrt{\mathscr{X}} \to \mathbb{C}, \quad \mu(X, g, Y) = \mathrm{Tr}(X \mathrm{Ad}_g Y).$$

Now we can define our main space where the convolution algebra dwells. The space $\sqrt{\mathscr{X}}$ has B-invariant projection to the first factor and our main space is the fibered product:

$$\mathscr{X} := \sqrt{\mathscr{X}} \times_{\mathfrak{g}} \sqrt{\mathscr{X}} = \mathfrak{g} \times G \times \mathfrak{n} \times G \times \mathfrak{n}.$$

The space $\mathscr{X}$ has a action of $G \times B^2$ that is induced from the $G \times B$ action on $\sqrt{\mathscr{X}}$, respectively the projections p_1, p_2 on two copies of $\sqrt{\mathscr{X}}$ are $G \times B^2$-equivariant. The group B is a semi-direct product $B = T \ltimes U$ of the torus T and the group of upper-triangular matrices U.

We define our main category to be:

$$\mathrm{MF}_n := \mathrm{MF}_{G \times B^2}(\mathscr{X}, W), \quad W = p_1^*(\mu) - p_2^*(\mu),$$

where we require the weak U^2-equivariance and strong $G \times T^2$-equivariance in our category. The strong $G \times T^2$-equivariance implies that all differentials in the complexes are $G \times T^2$-invariant. We can combine strong $G \times T^2$-equivariance with the weak U^2-equivariance since the Chevalley-Eilenberg complex for U^2 is $G \times T^2$-invariant.

There is an action of $T_{sc} = \mathbb{C}^* \times \mathbb{C}^* = \mathbb{C}_a^* \times \mathbb{C}_t^*$ on the space $\sqrt{\mathscr{X}}$ and the induced action on $\mathscr{X}$:

$$(\lambda, \mu) \cdot (X, g, Y) = (\lambda^2 \cdot X, g, \lambda^{-2}\mu^2 Y).$$

The potential W is not T_{sc}-invariant, it has weight 2 with respect to the torus $\mathbb{C}_t^*$. We require the differentials in a curved complex from MF_n to have weight 1 with respect to $\mathbb{C}_t^*$ and it has weight 0 with respect to $\mathbb{C}_a^*$. To simplify notations we do not use any extra indices to indicate such T_{sc}-equivariance. We also use notation

$$\mathbf{q}^k\mathbf{t}^l \cdot \mathscr{F}$$

for the matrix factorization $\mathscr{F}$ with the k-twisted $\mathbb{C}_a^*$-action and l-twisted $\mathbb{C}_t^*$-action.

Since the space $\mathscr{X}$ has B^2-action we can also twist a matrix factorization $\mathscr{F}$ by a representation of this group. Given a characters χ_l and χ_r of the left and right factor in B^2, the twisted matrix factorization is denoted by

$$\mathscr{F}\langle\chi_l, \chi_r\rangle.$$

To define convolution product in category MF_n we introduce the convolution space $\mathscr{X}_{con}$ which is a fibered product:

$$\mathscr{X}_{con} := \sqrt{\mathscr{X}} \times_{\mathfrak{g}} \sqrt{\mathscr{X}} \times_{\mathfrak{g}} \sqrt{\mathscr{X}} = \mathfrak{g} \times (G \times \mathfrak{n})^2.$$

There are three $G \times B^3$-equivariant maps $\pi_{12}, \pi_{23}, \pi_{13}$ and the convolution product is defined by the predictable formula:

$$\mathscr{F} \star \mathscr{G} := \pi_{13*}(\mathrm{CE}_{\mathfrak{n}^{(2)}}(\pi_{12}^*(\mathscr{F}) \otimes \pi_{23}^*(\mathscr{G}))^{T^{(2)}}).$$

Since the projections π_{ij} are smooth we can apply the base change formula. Hence the standard argument, that could be found in [3], implies that thus defined product is associative.

3.3.2 Knorrer Reduction

We can apply Knorrer periodicity discussed in Sect. 3.2.3 to reduce the size of our working space $\mathscr{X}$. Indeed, the pair of space and potential:

$$\bar{\mathscr{X}} = \mathfrak{b} \times G \times \mathfrak{n}, \quad \overline{W}(X, g, Y) = \mathrm{Tr}(X\mathrm{Ad}_g(Y))$$

is B^2-equivariant with respect to the action:

$$(b_1, b_2) \cdot (X, g, Y) = (\mathrm{Ad}_{b_1} X, b_1 g b_2, \mathrm{Ad}_{b_2}^{-1} Y).$$

Thus we can define the category of weakly U^2-equivariant and strongly T^2-equivariant matrix factorizations:

$$\overline{\mathrm{MF}}_n := \mathrm{MF}_{B^2}(\bar{\mathscr{X}}, \overline{W}).$$

To illustrate some of our methods we provide a proof for the equivalence in

Proposition 3.14 *There is an equivalence of categories:*

$$\Psi : \overline{\mathrm{MF}}_n \to \mathrm{MF}_n.$$

Proof First we observe that the group G acts freely on the space $\mathscr{X}$ hence we can take quotient by this group. The quotient can be implemented with help of the map:

$$\mathscr{X} \xrightarrow{q} \mathscr{X}^\circ := \mathfrak{g} \times \mathfrak{n} \times G \times \mathfrak{n}, \quad q(X, g_1, Y_1, g_2, Y_2) = (\mathrm{Ad}_{g_1}^{-1} X, Y_1, g_1^{-1} g_2, Y_2).$$

The potential $W^\circ(X, Y_1, g, Y_2) = \mathrm{Tr}(X(Y_1 - \mathrm{Ad}_g Y_2))$ is the pull-back $W_0 = q^*(W)$ and the pull-back provides an equivalence $q^* : \mathrm{MF}_n \simeq \mathrm{MF}_{B^2}(\mathscr{X}^0, W_0)$.

To complete our proof we fix notations for the truncation of a square matrix X:

$$X = X_+ + X_{--}, \quad X_+ \in \mathfrak{n}, \quad X_{--}^t \in \mathfrak{b}.$$

The potential W° can be written as a sum of $\overline{W}$ and a quadratic term and thus we can apply Knörrer periodicity:

$$\mathrm{Tr}(X(Y_1 - \mathrm{Ad}_g Y_2)) = \mathrm{Tr}((X_+ + X_{--})(Y_1 - \mathrm{Ad}_g Y_2)) = -\mathrm{Tr}(X_+(\mathrm{Ad}_g Y_2)) +$$
$$\mathrm{Tr}(X_{--}(Y_1 - \mathrm{Ad}_g Y_2)) = -\mathrm{Tr}(X_+(\mathrm{Ad}_g Y_2)) + \mathrm{Tr}(X_{--}(Y_1 - \mathrm{Ad}_g Y_2)_+).$$

The entries of matrices $X_{--}, Y_1 - (\mathrm{Ad}_g Y_2)_+$ are coordinates in the direction transversal to the subspace $\bar{\mathscr{X}}$ with coordinates X_+, g, Y_2 and Knorrer periodicity allows us to remove the quadratic term in the last formula. □

It is explained in [26] that the category $\overline{\mathrm{MF}}_n$ has a monoidal structure $\bar{\star}$ such that that the functor Ψ sends it to the monoidal structure $\star$.

3.3.3 *Braid Groups*

The affine braid group $\mathfrak{Br}_n^{aff}$ is the group of braids whose strands may also wrap around a 'flag pole'. The group is generated by the standard generators σ_i,

$i = 1, \ldots, n-1$ and a braid Δ_n that wraps the last stand of the braid around the flag pole:

$\sigma_i =$ 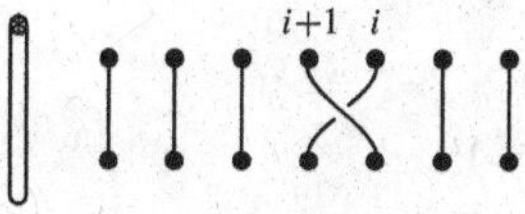

and $\Delta_n =$ 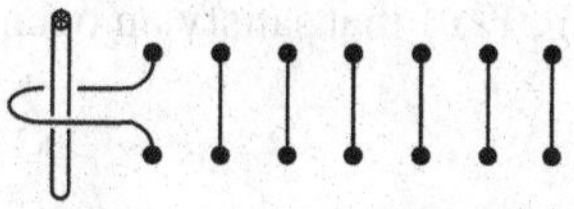 .

The defining relations for these generators are

$$\sigma_{n-1} \cdot \Delta_n \cdot \sigma_{n-1} \cdot \Delta_n = \Delta_n \cdot \sigma_{n-1} \cdot \Delta_n \cdot \sigma_{n-1},$$

$$\sigma_i \cdot \Delta_n = \Delta_n \cdot \sigma_i, \quad i < n-1,$$

$$\sigma_i \cdot \sigma_{i+1} \cdot \sigma_i = \sigma_{i+1} \cdot \sigma_i \cdot \sigma_{i+1}, \quad i = 1, \ldots, n-2,$$

$$\sigma_i \cdot \sigma_j = \sigma_j \cdot \sigma_i, \quad |i-j| > 1.$$

The mutually commuting Bernstein-Lusztig (BL) elements $\Delta_i \in \mathfrak{Br}_n^{aff}$ are defined as follows:

$\Delta_i = \sigma_i \cdots \sigma_{n-2}\sigma_{n-1}\Delta_n\sigma_{n-1}\sigma_{n-2}\cdots\sigma_i =$ 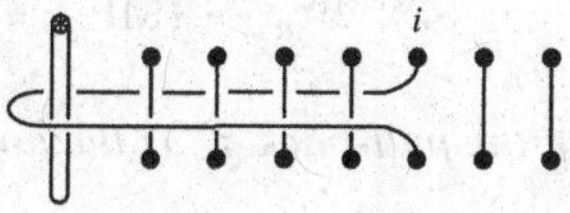

.

The finite braid group $\mathfrak{Br}_n$ is a subgroup of the affine braid group with the generators σ_i, $i = 1, \ldots, n-1$. Other words, we do not allow the braids to go around the pole.

There is a natural homomorphism $\mathfrak{fgt}\colon \mathfrak{Br}_n^{aff} \to \mathfrak{Br}_n$, geometrically it is defined by removing the flag pole. In particular we have:

$$\mathfrak{fgt}(\Delta_n) = 1, \quad \mathfrak{fgt}(\Delta_i) = \delta_i, \quad i = 1, \ldots, n-1.$$

The inclusion discussed above $i_{fin} : \mathfrak{Br}_n \to \mathfrak{Br}_n^{aff}$ is a section of the flag pole forgetting map: $\mathfrak{frg} \circ i_{fin} = 1$.

3.3.4 Braid Action

In this section we outline a construction of the homomorphisms from the (affine) braid group to our convolution algebras of matrix factorizations. For a geometric counter-part of the map $\mathfrak{fgt}$ we need to introduce *stable* versions of our categories of matrix factorizations.

Let us define the stable locus $\bar{\mathscr{X}}^{st,\bullet} \subset \bar{\mathscr{X}} \times V$ to be a set of quadruples (X, g, Y, v) that satisfy an open condition:

$$\mathbb{C}\langle(\mathrm{Ad}_g^{-1} X)_+, Y\rangle v = V. \tag{3.5}$$

There is a natural projection $\pi_V : \bar{\mathscr{X}} \times V \to \bar{\mathscr{X}}$ and there is an open embedding map $j_{st} : \bar{\mathscr{X}}^{st} \to \bar{\mathscr{X}}$ where $\bar{\mathscr{X}}^{st} = \pi_V(\bar{\mathscr{X}}^{\bullet,st})$. This map induces the pull-back map:

$$j_{st}^* : \overline{\mathrm{MF}}_n \to \overline{\mathrm{MF}}_n^{st} := \mathrm{MF}_{B^2}(\bar{\mathscr{X}}^{st}, \overline{W}).$$

It is shown in [26, Lemma 13.3] that the category $\overline{\mathrm{MF}}_n$ has a natural structure of convolution algebra. The main results of the papers [24, 26] play a crucial role in the construction of the knot invariant in the next section.

Theorem 3.15 *There are homomorphisms of algebras:*

$$\Phi : \mathfrak{Br}_n \to (\overline{\mathrm{MF}}_n^{st}, \bar{\star}), \quad \Phi^{aff} : \mathfrak{Br}_n^{aff} \to (\overline{\mathrm{MF}}_n, \bar{\star}).$$

Moreover, the pull-back j_{st}^* is the homomorphism of the convolution algebras and

$$j_{st}^* \circ \Phi^{aff} = \Phi \circ \mathrm{fgt}.$$

The fact that the pull-back morphism is an algebra homomorphism relies on the following shrinking lemma, for a proof see lemma 12.3 in [26].

Lemma 3.16 *Suppose X is a quasi-affine variety and $\mathscr{F} = (M, D) \in \mathrm{MF}(X, W)$, $W \in \mathbb{C}[X]$. The elements of $\mathbb{C}[X]$ act on $\mathrm{MF}(X, W)$ by multiplication. Let us assume that the elements of the ideal $I = (f_1, \dots, f_m) \subset \mathbb{C}[X]$ act by zero-homotopic endomorphisms on $\mathscr{F}$ and $Z' \subset X$ is the zero locus of I. Let $Z \subset X$ be a subvariety defined by $J = (g_1, \dots, g_n)$ such that $Z \cap Z' = \emptyset$. Then $\mathscr{F}$ is homotopic to $\mathscr{F}|_{X \setminus Z}$ as matrix factorization over $\mathbb{C}[X]$.*

In particular the lemma implies that we can shrink our ambient space to any open neighborhood of the critical locus of the potential and such operation does not change the corresponding category of matrix factorizations.

Let us also remark that there is another construction for the affine braid group action on the similar category of matrix factorizations in [2] but the precise relation between our construction and result of this paper is not known to the author.

3.4 Knot Invariants

3.4.1 Geometric Trace Operator

Let $\mathfrak{b}_n$, $\mathfrak{n}_n$ be Lie algebras of the group of upper, respectively strictly-upper triangular $n \times n$ matrices. The free nested Hilbert scheme FHilb_n^{free} is a $B \times \mathbb{C}^*$-quotient of the sublocus $\widetilde{\mathrm{FHilb}}_n^{free} \subset \mathfrak{b}_n \times \mathfrak{n}_n \times V_n$ of the cyclic triples

$$\widetilde{\mathrm{FHilb}}_n^{free} = \{(X, Y, v) | \mathbb{C}\langle X, Y\rangle v = V_n\},$$

here $V_n = \mathbb{C}^n$. The usual nested Hilbert scheme FHilb_n is a subvariety of FHilb_n^{free}, it is defined by the condition that X, Y commute.

Remark 3.17 There is a bit of confusion in the notations, what we denote here by FHilb_n is denoted in [26] by $\mathrm{Hilb}_{1,n}$ and by $\mathrm{FHilb}_n(\mathbb{C})$ in [11].

The torus $T_{sc} = \mathbb{C}^* \times \mathbb{C}^*$ acts on FHilb_n^{free} by scaling the matrices. We denote by $D_{T_{sc}}^{per}(\mathrm{FHilb}_n^{free})$ a derived category of the two-periodic complexes of the T_{sc}-equivariant quasi-coherent sheaves on FHilb_n^{free}. Let us also denote by $\mathscr{B}^\vee$ the descent of the trivial vector bundle V_n on $\widetilde{\mathrm{FHilb}}_n^{free}$ to the quotient FHilb_n^{free}. Respectively, $\mathscr{B}$ stands for the dual of $\mathscr{B}^\vee$. Below we construct for every $\beta \in \mathfrak{Br}_n$ an element

$$\mathbb{S}_\beta \in D_{T_{sc}}^{per}(\mathrm{FHilb}_n^{free})$$

such that space of hyper-cohomology of the complex:

$$\mathbb{H}^k(\mathbb{S}_\beta) := \mathbb{H}(\mathbb{S}_\beta \otimes \Lambda^k \mathscr{B})$$

defines an isotopy invariant.

Theorem 3.18 ([26]) *For any $\beta \in \mathfrak{Br}_n$ the doubly graded space*

$$H^k(\beta) := \mathbb{H}^{(k+\mathrm{writh}(\beta)-n-1)/2}(\mathbb{S}_\beta)$$

is an isotopy invariant of the braid closure $L(\beta)$.

The variety $\widetilde{\mathrm{FHilb}}_n^{free}$ embeds inside $\overline{\mathscr{X}}$ via a map $j_e : (X, Y, v) \to (X, e, Y, v)$. The diagonal copy $B = B_\Delta \hookrightarrow B^2$ respects the embedding j_e and since $j_e^*(\overline{W}) = 0$, we obtain a functor:

$$j_e^* : \mathrm{MF}_{B_n^2}(\overline{\mathscr{X}}^{st}, \overline{W}) = \overline{\mathrm{MF}}_n^{st} \to \mathrm{MF}_{B_\Delta}(\widetilde{\mathrm{FHilb}}_n^{free}, 0).$$

Respectively, we get a geometric version of "closure of the braid" map:

$$\mathbb{L} : \mathrm{MF}_{B_n^2}(\overline{\mathscr{X}}^{st}, \overline{W}) = \overline{\mathrm{MF}}_n^{st} \to D^{per}_{T_{sc}}(\mathrm{FHilb}_n^{free}).$$

The main result of [26] could be restated in more geometric terms via geometric trace map:

$$\mathscr{T}r : \mathfrak{Br}_n \to D^{per}_{T_{sc}}(\mathrm{FHilb}_n^{free}), \quad \mathscr{T}r(\beta) := \oplus_k \mathbb{L}(\Phi(\beta) \otimes \Lambda^k \mathscr{B}).$$

The above mentioned complex $\mathbb{S}_\beta$ is the complex $\mathbb{L}(\Phi(\beta))$. The differentials in the complex $\mathbb{S}_\beta$ are of degree t thus the differentials are invariant with respect to the anti-diagonal torus T_a. Hence the forgetful functor χ : $D^{per}_{T_{sc}}(\mathrm{FHilb}_n^{free}) \to D^{per}_{T_a}(\mathrm{FHilb}_n^{free})$ could be composed with K-theory functor $\mathrm{K} : D^{per}_{T_a}(\mathrm{FHilb}_n^{free}) \to K_{T_a}(\mathrm{FHilb}_n^{free})$. The composite functor $\mathrm{K} \circ \chi$ is closely related to decategorification and the classical Ocneanu-Jones trace, we discuss the Ocneaunu-Jones trace Tr^{OJ} and related theorem of Markov in the next subsection.

Theorem 3.19 ([26]) *The composition* $\mathbb{H} \circ \mathscr{T}r : \mathfrak{Br}_n \to D^{per}_{T_{sc}}(pt)$ *categorifies the Jones-Oceanu trace:*

$$\mathrm{Tr}^{OJ}(\beta) = \dim_{a,q} \mathrm{K} \circ \chi \circ \mathbb{H} \circ \mathscr{T}r(\beta),$$

where the q-grading comes from T_a*-action and a-grading from the exterior powers of* $\mathscr{B}$.

3.4.2 OJ Trace and Markov Theorem

As we discussed before every link L in $\mathbb{R}^3$ is isotopic to the closure of some braid $L = \mathrm{L}(\beta)$, $\beta \in \mathfrak{Br}_n$. On the other hand it is clear that such a presentation is not unique. Markov theorem describes the non-uniqueness explicitly and thus provides an algebraic description of the set $\mathfrak{L}$ of the isotopy equivalence classes of the links.

Theorem 3.20 *The closure operation* L *identifies the set* $\mathfrak{L}$ *of isotopy class of links in* S^3 *and the set of equivalence classes:*

$$\bigcup_n \mathfrak{Br}_n / \sim$$

where the equivalence relation is generated by the elementary equivalences:

$$\alpha \cdot \beta \sim \beta \cdot \alpha, \quad \alpha, \beta \in \mathfrak{Br}_n \tag{3.6}$$

$$\mathfrak{Br}_{n+1} \ni \alpha \cdot \sigma^{\pm 1} \sim \alpha, \quad \alpha \in \mathfrak{Br}_n. \tag{3.7}$$

If we have homomorphism Tr from the braid group to some field F that respects the relations (3.7) then the value $\mathrm{Tr}(\beta) \in F$ is an isotopy invariant of the closure $\mathrm{L}(\beta)$. In practice it is hard to classify such homomorphisms, however the great discovery of Ocneanu and Jones is that one can classify such homomorphisms if we pass to a quotient H_n of the braid group.

The Hecke algebra H_n is generated by g_i, $i = 1, \ldots, n-1$ modulo relations:

$$g_i g_{i+1} g_i = g_{i+1} g_i g_{i+1}, \quad i = 1, \ldots, n-2,$$

$$g_i - g_i^{-1} = q - q^{-1}, \quad i = 1, \ldots, n-1.$$

There is a natural algebra homomorphism $\pi : \mathfrak{Br}_n \to H_n$, $\sigma_i \mapsto g_i$. It is shown in [15] that there is a unique homomorphism $\mathrm{Tr}^{OJ} : \bigcup_n H_n \to \mathbb{Q}(a, q)$ that satisfies relations (3.7) and normalizing relation

$$\mathrm{Tr}^{OJ}(\mathbf{1}) = \frac{a - a^{-1}}{q - q^{-1}}.$$

The corresponding invariant is $\mathrm{Tr}^{OJ}(\beta) \in \mathbb{Q}(a, q)$ is also known as HOMFLYPT invariant, HOMFLYPT(β).

Thus the formula (3.1) from the introduction and Theorem 3.19 state that there is a specialization of the graded dimension of HHH(β) that becomes a HOMFLYPT invariant. Let us recall that the space HHH(β) up to some elementary grading shift is equal:

$$\mathbb{H}^*(j_e^*(\Phi(\beta) \otimes \Lambda^\bullet \mathscr{B})^B).$$

This space naturally has four gradings: $*$, $\bullet$ and T_{sc}-grading. However, only three of these gradings are invariant with respect to the Markov moves: $*$ is not preserved by the moves. The first grading is $\bullet$, we call it a-grading, since it is responsible for the a-variable in the HOMFLYPT polynomial specializations. The other two gradings come from T_{sc}-action:

$$\deg(X_{ij}) = q^2, \quad \deg(Y_{ij}) = q^{-2}t^{-2}.$$

To specialize to the HOMFLYPT polynomial we need to set $t = -1$ or more geometrically, we need to restrict the torus T_{sc}-action on the space HHH(β) to the action of the anti-diagonal torus, we denote the specialized category by $\overline{\mathrm{MF}}_n^{st,a}$. To be more precise, the category $\mathrm{MF}_n^{st,a}$ is a category of matrix factorizations on $\overline{\mathscr{X}}^{st}$ with the potential $\overline{W}$ which are B_n^2-weakly equivariant, $G \times T^2 \times T_a$-strongly equivariant.

As we mentioned before this torus is special because under this specialization the differentials in the curved complexes from $\overline{\mathrm{MF}}_n^{st}$ become torus invariant, hence there is a well-defined functor:

$$\mathrm{K}: \quad \overline{\mathrm{MF}}_n^{st,a} \to \mathrm{K}_{T_a}(\mathrm{MF}_n^{st,a}).$$

This K-theory functor turns the homotopy equivalence (3.10) into the relation:

$$[\mathscr{C}_+] = q^{-1}([\mathscr{C}_{\|}] - [\mathscr{C}_{\bullet}\langle -\chi_1, -\chi_1\rangle]).$$

Thus the combination of the relations (3.2) and (3.12), (3.13) imply the quadratic relation in the Hecke algebra.

The Markov move relations (3.7) hold for the invariant HHH(β) and in Sect. 3.7 we discuss the main idea of our proof of the Markov moves for HHH. We need some details of the braid group action construction for the Markov move argument, therefore we outline the construction in the next section.

3.5 Geometric Realization of the Affine Braid Group

3.5.1 Induction Functors

The standard parabolic subgroup P_k has Lie algebra generated by $\mathfrak{b}$ and $E_{i+1,i}$, $i \neq k$. Let us define space $\overline{\mathscr{X}}(P_k) := \mathfrak{b} \times P_k \times \mathfrak{n}$ and let us also use notation $\overline{\mathscr{X}}(\mathrm{GL}_n)$ for $\overline{\mathscr{X}}$. There is a natural embedding $\bar{i}_k : \overline{\mathscr{X}}(P_k) \to \overline{\mathscr{X}}$ and a natural projection $\bar{p}_k : \overline{\mathscr{X}}(P_k) \to \overline{\mathscr{X}}(\mathrm{GL}_k) \times \overline{\mathscr{X}}(\mathrm{GL}_{n-k})$. The space $\overline{\mathscr{X}}(\mathrm{GL}_k) \times \overline{\mathscr{X}}(\mathrm{GL}_{n-k})$ is equipped with a $B_k^2 \times B_{n-k}^2$-invariant potential $\overline{W}^{(1)} + \overline{W}^{(2)}$ which is a sum of pull-backs of the potentials W along the projection on the first and the second factors. Moreover, we have:

$$\bar{i}_k^*(\overline{W}) = \bar{p}_k^*(\overline{W}^{(1)} + \overline{W}^{(2)}). \tag{3.8}$$

Since the embedding $\bar{i}_k$ satisfies the conditions for existence of the push-forward and the relation (3.8) between the potentials holds, we can define the induction functor:

$$\overline{\mathrm{ind}}_k := \bar{i}_{k*} \circ \bar{p}_k^* : \mathrm{MF}_{B_k^2}(\overline{\mathscr{X}}(\mathrm{GL}_k), \overline{W}) \times \mathrm{MF}_{B_{n-k}^2}(\overline{\mathscr{X}}(\mathrm{GL}_{n-k}), \overline{W})$$
$$\to \mathrm{MF}_{B_n^2}(\overline{\mathscr{X}}(\mathrm{GL}_n), \overline{W})$$

Similarly we define space $\overline{\mathscr{X}}^{st}(P_k) \subset \mathfrak{b} \times P_k \times \mathfrak{n} \times V$ as an open subset defined by the stability condition (3.5). The last space has a natural projection map

$$\bar{p}_k : \overline{\mathscr{X}}^{st}(P_k) \to \overline{\mathscr{X}}(\mathrm{GL}_k) \times \overline{\mathscr{X}}^{st}(\mathrm{GL}_{n-k})$$

and the embedding $\bar{i}_k : \overline{\mathscr{X}}^{st}(P_k) \to \overline{\mathscr{X}}^{st}(\mathrm{GL}_n)$ and we can define the induction functor:

$$\overline{\mathrm{ind}}_k := \bar{i}_{k*} \circ \bar{p}_k^* : \mathrm{MF}_{B_k^2}(\overline{\mathscr{X}}(\mathrm{GL}_k), \overline{W}) \times \mathrm{MF}_{B_{n-k}^2}(\overline{\mathscr{X}}^{st}(\mathrm{GL}_{n-k}), \overline{W})$$
$$\to \mathrm{MF}_{B_n^2}(\overline{\mathscr{X}}^{st}(\mathrm{GL}_n), \overline{W})$$

It is shown in section 6 (proposition 6.2) of [26] that the functor $\overline{\mathrm{ind}}_k$ is a homomorphism of the convolution algebras:

$$\overline{\mathrm{ind}}_k(\mathscr{F}_1 \boxtimes \mathscr{F}_2)\bar{\star}\overline{\mathrm{ind}}_k(\mathscr{G}_1 \boxtimes \mathscr{G}_2) = \overline{\mathrm{ind}}_k(\mathscr{F}_1\bar{\star}\mathscr{G}_2 \boxtimes \mathscr{F}_2\bar{\star}\mathscr{G}_2).$$

To define the non-reduced version of the induction functors one needs to introduce the space $\mathscr{X}^\circ(\mathrm{GL}_n) = \mathfrak{g} \times \mathrm{GL}_n \times \mathfrak{n} \times \mathfrak{n}$ which is the slice to GL_n-action on the space $\mathscr{X}$. In particular, the potential W on this slice becomes:

$$W(X, g, Y_1, Y_2) = \mathrm{Tr}(X(Y_1 - \mathrm{Ad}_g(Y_2))).$$

Similarly to the case of the reduced space, one can define the space $\mathscr{X}^\circ(P_k) := \mathfrak{g} \times P_k \times \mathfrak{n} \times \mathfrak{n}$ to be a subvariety of $\mathscr{X}^\circ(\mathrm{GL}_n)$ and the corresponding maps $i_k : \mathscr{X}^\circ(P_k) \to \mathscr{X}^\circ(\mathrm{GL}_n)$, $p_k : \mathscr{X}^\circ(P_k) \to \mathscr{X}^\circ(\mathrm{GL}_k) \times \mathscr{X}^\circ(\mathrm{GL}_{n-k})$. Thus we get a version of the induction functor for non-reduced spaces:

$$\mathrm{ind}_k := i_{k*} \circ p_k^* : \mathrm{MF}_{B_k^2}(\mathscr{X}(\mathrm{GL}_k), W) \times \mathrm{MF}_{B_{n-k}^2}(\mathscr{X}(\mathrm{GL}_{n-k}), W)$$
$$\to \mathrm{MF}_{B_n^2}(\mathscr{X}(\mathrm{GL}_n), W)$$

It is shown in proposition 6.1 of [26] that the Knörrer functor is compatible with the induction functor:

$$\mathrm{ind}_k \circ (\Phi_k \times \Phi_{n-k}) = \Phi_n \circ \mathrm{ind}_k.$$

3.5.2 Generators of the Finite Braid Group Action

Let us define B^2-equivariant embedding $i : \overline{\mathscr{X}}(B_n) \to \overline{\mathscr{X}}, \overline{\mathscr{X}}(B) := \mathfrak{b} \times B \times \mathfrak{n}$. The pull-back of $\overline{W}$ along the map i vanishes and the embedding i satisfies the conditions for existence of the push-forward $i_* : \mathrm{MF}_{B^2}(\overline{\mathscr{X}}(B_n), 0) \to \mathrm{MF}_{B^2}(\overline{\mathscr{X}}(\mathrm{GL}_n), \overline{W})$. We denote by $\mathbb{C}[\overline{\mathscr{X}}(B_n)] \in \mathrm{MF}_{B^2}(\overline{\mathscr{X}}(B_n), 0)$ the matrix factorization with zero differential that is concentrated only in even homological degree. As it is shown in proposition 7.1 of [26] the push-forward

$$\mathbb{1}_n := i_*(\mathbb{C}[\overline{\mathscr{X}}(B_n)])$$

is the unit in the convolution algebra. Similarly, $\mathbb{1}_n := \Phi(\bar{\mathbb{1}}_n)$ is also a unit in non-reduced case.

Let us first discuss the case of the braids on two strands. The key to our construction of the braid group action in [26] is the following factorization in the case $n = 2$:

$$\overline{W}(X, g, Y) = y_{12}(2g_{11}x_{11} + g_{21}x_{12})g_{21}/\det,$$

where $\det = \det(g)$ and

$$g = \begin{bmatrix} g_{11} & g_{12} \\ g_{21} & g_{22} \end{bmatrix}, \quad X = \begin{bmatrix} x_{11} & x_{12} \\ 0 & x_{22} \end{bmatrix}, \quad Y = \begin{bmatrix} 0 & y_{12} \\ 0 & 0 \end{bmatrix}$$

Thus we can define the following strongly equivariant Koszul matrix factorization:

$$\bar{\mathscr{C}}_+ := (\mathbb{C}[\overline{\mathscr{X}}] \otimes \Lambda\langle\theta\rangle, D) \in \mathrm{MF}^{str}_{B^2}(\overline{\mathscr{X}}, \overline{W}),$$

$$D = \frac{g_{12}y_{12}}{\det}\theta + \left(g_{11}(x_{11} - x_{22}) + g_{21}x_{12}\right)\frac{\partial}{\partial\theta},$$

where $\Lambda\langle\theta\rangle$ is the exterior algebra with one generator.

This matrix factorization corresponds to the positive elementary braid on two strands. Using the induction functor we can extend the previous definition to the case of the arbitrary number of strands. For that we introduce an insertion functor:

$$\overline{\mathrm{Ind}}_{k,k+1} : \mathrm{MF}_{B_2^2}(\overline{\mathscr{X}}(\mathrm{GL}_2), \overline{W}) \to \mathrm{MF}_{B_n^2}(\overline{\mathscr{X}}(\mathrm{GL}_n), \overline{W})$$

$$\overline{\mathrm{Ind}}_{k,k+1}(\mathscr{F}) := \overline{\mathrm{ind}}_{k+1}(\overline{\mathrm{ind}}_{k-1}(\bar{\mathbb{1}}_{k-1} \times \mathscr{F}) \times \bar{\mathbb{1}}_{n-k-1}),$$

and similarly we define non-reduced insertion functor

$$\mathrm{Ind}_{k,k+1} : \mathrm{MF}_{B_2^2}(\mathscr{X}(G_2), W) \to \mathrm{MF}_{B_n^2}(\mathscr{X}(G_n), W).$$

Thus we define the generators of the braid group as follows:

$$\bar{\mathscr{C}}_+^{(k)} := \overline{\mathrm{Ind}}_{k,k+1}(\overline{\mathrm{ind}}_{k-1}(\bar{\mathscr{C}}_+)), \quad \mathscr{C}_+^{(k)} := \mathrm{Ind}_{k,k+1}(\mathrm{ind}_{k-1}(\mathscr{C}_+)).$$

Section 11 of [26] is devoted to the proof of the braid relations between these elements:

$$\bar{\mathscr{C}}_+^{(k+1)} \bar{\star} \bar{\mathscr{C}}_+^{(k)} \bar{\star} \bar{\mathscr{C}}_+^{(k+1)} = \bar{\mathscr{C}}_+^{(k)} \bar{\star} \bar{\mathscr{C}}_+^{(k+1)} \bar{\star} \bar{\mathscr{C}}_+^{(k)},$$

$$\mathscr{C}_+^{(k+1)} \star \mathscr{C}_+^{(k)} \star \mathscr{C}_+^{(k+1)} = \mathscr{C}_+^{(k)} \star \mathscr{C}_+^{(k+1)} \star \mathscr{C}_+^{(k)}.$$

Let us now discuss the inversion of the elementary braid. In view of inductive definition of the braid group action, it is sufficient to understand the inversion in the case $n = 2$.

Thus we define:

$$\mathscr{C}_- := \mathscr{C}_+\langle -\chi_1, \chi_2\rangle \in \mathrm{MF}_{B^2}(\mathscr{X}(\mathrm{GL}_2), W),$$

where χ_1, χ_2 are the standard generators of the character group of $\mathbb{C}^* \times \mathbb{C}^* = T \subset B_2$. The definition of $\bar{\mathscr{C}_-}$ is similar. It could be shown that the definition of $\mathscr{C}_-$ is actually symmetric with respect to the left-right twisting:

$$\mathscr{C}_- = \mathscr{C}_+\langle \chi_2, -\chi_1\rangle.$$

Theorem 3.21 ([26]) *We have:*

$$\mathscr{C}_+ \star \mathscr{C}_- = \mathbb{1}_2. \tag{3.9}$$

3.6 Sample Computation

In this section we would like to show an example of the convolution algebra computations. But before we would expand a little bit our discussion of the basic matrix factorizations in the case of $n = 2$.

3.6.1 *Basic Matrix Factorizations of Rank 2*

We have shown in the previous section that the potential $\overline{W}$ is a product of three factors and we used this fact to define the matrix factorization $\mathscr{C}_+$. However, it is clear that there are two more natural matrix factorizations for this potential:

$$\bar{\mathscr{C}_\parallel} := (\mathbb{C}[\overline{\mathscr{X}}]\otimes\Lambda\langle\theta\rangle, D_\parallel, 0, 0), \quad \bar{\mathscr{C}_\bullet} := (\mathbb{C}[\overline{\mathscr{X}}]\otimes\Lambda\langle\theta\rangle, D_\bullet, 0, 0) \in \mathrm{MF}_{B^2}(\overline{\mathscr{X}}, \overline{W}),$$

$$D_\parallel = \frac{g_{21}}{\det}\theta + y_{12}\tilde{x}_0\frac{\partial}{\partial\theta}, \quad D_\bullet = \frac{g_{21}}{\det}\tilde{x}_0\theta + y_{12}\frac{\partial}{\partial\theta}, \quad \tilde{x}_0 = g_{11}(x_{11}-x_{22}) + g_{21}x_{12}.$$

One of the matrix factorizations is actually a cone of the morphism between the other two:

$$[\bar{\mathscr{C}_\parallel} \xrightarrow{\phi} \bar{\mathscr{C}_\bullet}\langle -\chi_1, -\chi_1\rangle] \sim \mathbf{qt}\cdot\bar{\mathscr{C}_+} \tag{3.10}$$

with map ϕ defined by

$$\begin{array}{ccc} \mathrm{t}^{-1}\cdot R\langle\chi_2,-\chi_1\rangle & \xrightarrow{\quad 1 \quad} & R\langle\chi_2,-\chi_1\rangle \\ {\scriptstyle y_{12}\tilde{x}_0}\uparrow\downarrow{\scriptstyle g_{21}} & & {\scriptstyle y_{12}}\uparrow\downarrow{\scriptstyle g_{21}\tilde{x}_0} \\ R & \xrightarrow{\quad \tilde{x}_0 \quad} & R\langle 0,-\chi_1\rangle, \end{array}$$

where $R=\mathbb{C}[\overline{\mathscr{X}}]$. This relation is crucial for our discussion of the connection with the Oceanus-Jones traces

3.6.2 Details on the Convolution Product

The convolution product inside the category $\mathrm{MF}_{B^2}(\overline{\mathscr{X}},\overline{W})$ is a bit tricky to define and we refer reader to our paper [26] where the convolution product is constructed and used for the computations for $n=3$. On the other hand the space $\mathscr{X}$ is bigger than the space $\overline{\mathscr{X}}$ but the construction of the convolution is more straightforward. The space $\mathscr{X}^\circ := \mathscr{X}/\mathrm{GL}_n = \mathfrak{g}\times\mathfrak{n}\times\mathrm{GL}_n\times\mathfrak{n}$ is intermediate between these two spaces and we choose to work with this slightly bigger space to make our exposition simpler.

The space $\mathscr{X}^\circ$ and the relevant potential W° appeared already in the proof of Proposition 3.14. Let us spell out the definition of the convolution structure for elements $\mathscr{F},\mathscr{G}\in\mathrm{MF}_{B^2}(\mathscr{X}^\circ,W^\circ)$:

$$\mathscr{F}\star\mathscr{G} := \pi^\circ_{13*}(\mathrm{CE}(\pi^{\circ*}_{12}(\mathscr{F})\otimes\pi^{\circ*}_{23}(\mathscr{G}))),$$

where we used the convolution space $\mathscr{X}^\circ_{cnv} := \mathfrak{g}\times\mathfrak{n}\times\mathrm{GL}_n\times\mathfrak{n}\times\mathrm{GL}_n\times\mathfrak{n}$ and the B^3-equivariant maps are

$$\pi^\circ_{12}(X,Y_1,g_{12},Y_2,g_{23},Y_3)=(X,Y_1,g_{12},Y_2),$$

$$\pi^\circ_{23}(X,Y_1,g_{12},Y_2,g_{23},Y_3)=(\mathrm{Ad}_{g_{12}}X,Y_2,g_{23},Y_3),$$

$$\pi^\circ_{13}(X,Y_1,g_{12},Y_2,g_{23},Y_3)=(X,Y_1,g_{12}g_{23}^{-1},Y_3).$$

To write the versions $\mathscr{C}_\parallel,\mathscr{C}_\bullet,\mathscr{C}_+$ of the matrix factorizations from above we need more precise notations for the Koszul matrix factorizations. We use the matrix notation

$$\begin{bmatrix} a_1 & b_1 & \theta_1 \\ \vdots & \vdots & \vdots \\ a_m & b_m & \theta_m \end{bmatrix}.$$

for the matrix factorization from MF(X, F) with the differential $D = \sum_{i=1}^m a_i\theta_i + b_i\frac{\partial}{\partial\theta_i}$ acting on $\mathbb{C}[X] \otimes \Lambda^\bullet[\theta]$.

Let us also fix coordinates on the space $\mathscr{X}^\circ = \mathfrak{g} \times \mathfrak{b} \times G \times \mathfrak{b}$:

$$X = \begin{bmatrix} x_0 + \mathrm{tr}/2 & x_1 \\ x_{-1} & -x_0 + \mathrm{tr}/2 \end{bmatrix}, \quad Y_i = \begin{bmatrix} 0 & y_i \\ 0 & 0 \end{bmatrix} \quad g = \begin{bmatrix} a_{11} & a_{12} \\ a_{21} & a_{22} \end{bmatrix},$$

where tr $=$ tr$X$. We also denote by δ_1, δ_2 the generators of Lie (U^2), $U^2 \subset B^2$. We also only indicate non-trivial actions of δ_i, that is if no action of δ_i is given then this action is trivial.

With this conventions we have the matrix factorization of the identity braid has the form

$$\mathscr{C}_{\|} = \begin{bmatrix} x_{-1} & y_1 - y_2 a_{11}^2 & \theta_1 \\ y_2\tilde{x}_0 & a_{21} & \theta_2 \end{bmatrix}, \quad \delta_1\theta_1 = -2y_2a_{11}\theta_2.$$

The blob matrix factorization has the form

$$\mathscr{C}_\bullet = \begin{bmatrix} x_{-1} & y_1 - y_2 a_{11}^2 & \theta_1 \\ a_{21}\tilde{x}_0 & y_2 & \theta_2' \end{bmatrix}, \quad \delta_1\theta_1 = -2a_{21}a_{11}\theta_2'$$

or equivalently

$$\mathscr{C}_\bullet = \begin{bmatrix} x_{-1} & y_1 & \theta_1' \\ -a_{11}^2 x_{-1} + a_{21}\tilde{x}_0 & y_2 & \theta_2' \end{bmatrix}, \quad \theta_1' = \theta_1 + a_{22}^2\theta_2', \quad \delta_1\theta_1' = 0$$

The matrix factorization of the positive intersection is

$$\mathscr{C}_+ = \begin{bmatrix} x_{-1} & y_1 - y_2 a_{11}^2 & \theta_1 \\ \tilde{x}_0 & a_{21}y_2 & \theta_2 \end{bmatrix}, \quad \delta_1\theta_1 = -2a_{11}\theta_2. \qquad (3.11)$$

3.6.3 Computation

Now we are ready to do our sample computation.

Proposition 3.22 *In the convolution algebra of* MF$_{\mathrm{GL}_n \times B^2}(\mathscr{X}, W)$ *we have:*

$$\mathscr{C}_\bullet\langle 0, \chi_1\rangle \star \mathscr{C}_\bullet\langle 0, \chi_1\rangle = \mathscr{C}_\bullet\langle \chi_1, \chi_1\rangle \oplus \mathscr{C}_\bullet\langle \chi_2, \chi_1\rangle.$$

Proof Let us fix some notation for the coordinates on the spaces that appear in our constructions. For the group elements in the product $\mathscr{X}^\circ_{conv} = \mathfrak{g} \times \mathfrak{n} \times \mathrm{GL}_2 \times \mathfrak{n} \times \mathrm{GL}_2 \times \mathfrak{n}$ we use notations a, b and for the non-zero elements of upper-triangular matrices in the product we use y_1, y_2, y_3. We also add prime to the conjugate of X: $X' = \mathrm{Ad}_a X$.

Thus the matrix factorization $\mathscr{C}'' = \pi_{12}^{\circ *}(\mathscr{C}_\bullet) \otimes \pi_{23}^{\circ *}(\mathscr{C}_\bullet)$ is the following Koszul matrix factorization:

$$\mathscr{C}'' = \begin{bmatrix} x_{-1} & y_1 - y_2 a_{11}^2 & \theta_1 \\ a_{21}(2x_0 a_{11} + x_1 a_{21}) & y_2 & \theta_2 \\ x'_{-1} & y_2 - y_3 b_{11}^2 & \theta_3 \\ b_{21}(2x'_0 b_{11} + x'_1 b_{21}) & y_3 & \theta_4 \end{bmatrix}, \quad \delta_1\theta_1 = -2a_{11}\theta_2, \quad \delta_2\theta_3 = -2b_{11}\theta_4.$$

By making suitable linear change of $\theta_1 \mapsto \theta_1 + 2a_{11}\theta_2, \theta_2 \mapsto \theta_2$ and $\theta_3 \mapsto \theta_3 + b_{11}\theta_4, \theta_4 \mapsto \theta_4$ we can make the first simplification of this matrix factorization:

$$\mathscr{C}'' = \begin{bmatrix} x_{-1} & y_1 & \theta_1 \\ -a_{11}^2 x_{-1} + a_{21}(2x_0 a_{11} + x_1 a_{21}) & y_2 & \theta_2 \\ x'_{-1} & y_2 & \theta_3 \\ -b_{11}^2 x'_{-1} + b_{21}(2x'_0 b_{11} + x'_1 b_{21}) & y_3 & \theta_4 \end{bmatrix}, \qquad \delta_i\theta_j = 0.$$

We use the third row to remove y_2 from the other rows:

$$\mathscr{C}'' = \begin{bmatrix} x_{-1} & y_1 & \theta_1 \\ -a_{11}^2 x_{-1} + a_{21}(2x_0 a_{11} + x_1 a_{21}) & 0 & \theta'_2 \\ 0 & y_2 & \theta_3 \\ -b_{11}^2 x'_{-1} + b_{21}(2x'_0 b_{11} + x'_1 b_{21}) & y_3 & \theta_4 \end{bmatrix}, \qquad \theta'_2 = \theta_2 - \theta_3.$$

Since θ_3 is B^2 invariant element, we can now remove the third row altogether and work over the ring $R' = \mathbb{C}[\mathscr{X}^\circ_{conv}]/(y_2)$.

We can also use the relation

$$-b_{11}^2 x'_{-1} + b_{21}(2x'_0 b_{11} + x'_1 b_{21}) = -c_{11}^2 x_{-1} + c_{21}(2x_0 c_{11} + x_1 c_{21})$$

to arrive to

$$\mathscr{C}_\bullet \star \mathscr{C}_\bullet = \begin{bmatrix} x_{-1} & y_1 & \theta_1 \\ -a_{11}^2 x_{-1} + a_{21}(2x_0 a_{11} + x_1 a_{21}) & 0 & \theta'_2 \\ -c_{11}^2 x_{-1} + c_{21}(2x_0 c_{11} + x_1 c_{21}) & y_3 & \theta_4 \end{bmatrix}$$

Doing couple more simple row transformations, that change the basis in the space $\langle \theta_1, \theta'_2, \theta'_4 \rangle$, we arrive to a simplified presentation of $\mathscr{C}_\bullet \star \mathscr{C}_\bullet$:

$$\mathscr{C}'' = \begin{bmatrix} x_{-1} & y_1 - c_{11}^2 y_3 & \theta'_1 \\ a_{21}(2x_0 a_{11} + x_1 a_{21}) & 0 & \theta''_2 \\ c_{21}(2x_0 c_{11} + x_1 c_{21}) & y_3 & \theta'_4 \end{bmatrix}$$

Now let us notice that the top and the bottom lines of the last Koszul complex are δ_2-invariant and together they form a Koszul matrix factorization $\pi_{13}^{\circ,*}(\mathscr{C}_\bullet)$. On the other hand the middle line has only one non-trivial differential and to complete the proof we need to compute the Chevalley-Eilenberg homology

$$H^*_{\mathrm{Lie}}(\mathfrak{n}, R'' \xrightarrow{f} R'')^T, \quad f = -a_{21}(2x_0a_{11} + x_1a_{21}),$$

where $R' = R'' \otimes \mathbb{C}[\mathrm{GL}_2]$ with last copy of GL_2 has coordinates c_{ij}.

The space $Spec(R'')$ has coordinates a, y_1, y_2, x and the Lie algebra $\mathfrak{n}$ only acts on the entries of the matrix a:

$$\delta_2 a_{i2} = -a_{i1}, \quad \delta_2 a_{i1} = 0.$$

The differential in complex for H^*_{Lie} is exactly δ_2 hence

$$H^0(\mathfrak{n}, R'') = \mathbb{C}[y_1, y_2, x, a_{11}, a_{21}, \det^{\pm 1}],$$
$$H^1(\mathfrak{n}, R'') = \mathbb{C}[y_1, y_2, x, a, \det^{\pm 1}]/(a_{11}, a_{21}),$$

where $\det = \det a$. Now we can extract the torus invariant part:

$$(H^*(\mathfrak{n}, R'') \otimes \chi_1)^{T_{sc}} = (H^0(\mathfrak{n}, R'') \otimes \chi_1)^{T_{sc}} = \langle a_{11}, a_{21} \rangle.$$

Finally, let us observe that the function f is quadratic on a hence its induced action on $(H^*(\mathfrak{n}, R'') \otimes \chi_1)^{T_{sc}}$ is trivial and the statement follows. □

Now let us derive the formula (3.2) from the above computation. For that let us recall that the stable locus $\bar{\mathscr{X}}^{st}$ is a union of two open subsets: $U_y = \{y \neq 0\}$, $U_x = \{(\mathrm{Ad}_g^{-1} X)_{12} \neq 0\}$. On the open set U_y the matrix factorization $\bar{\mathscr{C}}_\bullet$ contracts since y is one of the differentials of the curved complexes. Thus we can safely restrict our attention to the open locus U_x but on this locus $(\mathrm{Ad}_g^{-1} X)_{12} \neq 0$. Since the weights of $T_{sc} \times B^2$ on this non-vanishing elements are:

$$\mathrm{weight}((\mathrm{Ad}_g^{-1} X)_{12}) = \mathbf{q}^2 \cdot \langle 0, -\chi_1 + \chi_2 \rangle, \quad \mathrm{weight}(\det(a)) = \langle \chi_1 + \chi_2, \chi_1 + \chi_2 \rangle$$

we can trade the Borel action weight for $\mathbf{q}$ χ_1-shifts for $\mathbf{q}$-shifts:

$$\bar{\mathscr{C}}_\bullet \langle \chi' + \chi_1, \chi'' \rangle = \mathbf{q}^2 \bar{\mathscr{C}}_\bullet \langle \chi' - \chi_2, \chi'' - 2\chi_2 \rangle, \tag{3.12}$$

$$\bar{\mathscr{C}}_\bullet \langle \chi', \chi' + \chi_1 \rangle = \mathbf{q}^{-2} \bar{\mathscr{C}}_\bullet \langle \chi', \chi'' + \chi_2 \rangle. \tag{3.13}$$

Finally, we refer to theorem 3.33 that implies that the pull-back j^*_{st} turns the shifts $\langle \chi_2, 0 \rangle$ and $\langle 0, \chi_2 \rangle$ to the trivial B^2-equivariant shift.

3.7 Markov Relations

The first Markov relation is equivalent to HHH being a trace, that is we need to show that the functor HHH is constant on the conjugacy classes inside $\mathfrak{Br}_n$. In fact one can show a stronger statement. Before we state this stronger statement let us discuss the connection with usual flag Hilbert schemes.

3.7.1 Sheaves on the Flag Hilbert Scheme

The usual flag Hilbert scheme FHilb_n is a subvariety of FHilb_n^{free} defined by the commutativity constraint on X, Y:

$$[X, Y] = 0.$$

It turns out that the support of the homology of the complex $\mathbb{S}_\beta$ is contained in FHilb_n. Hence the sheaf homology of the complex is the sheaf

$$\mathscr{S}_\beta = \mathscr{S}_\beta^{odd} \oplus \mathscr{S}_\beta^{even} := \mathscr{H}^*(\mathrm{FHilb}_n^{free}, \mathbb{S}_\beta)$$

on $\mathrm{Hilb}_{1,n}$ and we immediately have the following:

Theorem 3.23 *There is a spectral sequence with E_2 term being*

$$(\mathrm{H}^*(\mathrm{FHilb}_n, \mathscr{S}_\beta \otimes \Lambda^k \mathscr{B}), d)$$

$$d : \mathrm{H}^k(\mathrm{FHilb}_n, \mathscr{S}_\beta^{odd/even} \otimes \Lambda^k \mathscr{B}) \to \mathrm{H}^{k-1}(\mathrm{FHilb}_n, \mathscr{S}_\beta^{even/odd} \otimes \Lambda^k \mathscr{B}),$$

that converges to $\mathbb{H}^k(\beta)$.

The theorem follows almost immediately from the main theorem 3.18 and the proposition 3.5. Moreover the sheaf $\mathscr{S}_\beta$ is actually is a conjugacy invariant:

Theorem 3.24 ([26]) *For any $\alpha, \beta \in \mathfrak{Br}_n$ we have:*

$$\mathscr{S}_{\alpha\cdot\beta} \simeq \mathscr{S}_{\beta\cdot\alpha}.$$

The argument could be found in the cited paper, here we illustrate the idea by showing that

$$\mathscr{S}_{\sigma_i\sigma_j\sigma_k} \simeq \mathscr{S}_{\sigma_j\sigma_k\sigma_i}. \tag{3.14}$$

Indeed, let us introduce the space $\mathscr{X}_3 \subset \mathfrak{gl}_n \times (\mathrm{GL}_n \times \mathfrak{n}_n)^3$ defined by the constraint requiring the cyclic product of the group elements to be one. There is a natural B^3-

action and B^3-equivariant projections:

$$pr_i : \mathscr{X}_3 \to \mathfrak{gl}_n \times \mathfrak{n}_n, \quad pr_i(X, g_1, Y_1, g_2, Y_2, g_3, Y_3) = (X, Y_i).$$

Respectively, we also have projections $\pi^\circ_{12}, \pi^\circ_{23}, \pi^\circ_{31} : \mathscr{X}_3 \to \mathscr{X}^\circ$ and

$$\mathbb{S}_{\sigma_i\sigma_j\sigma_k} = pr_{1*}(\mathscr{C}), \quad \mathbb{S}_{\sigma_j\sigma_k\sigma_i} = pr_{2*}(\mathscr{C}),$$
$$\mathscr{C} = \mathrm{CE}_{\mathfrak{n}^3}(\pi^{\circ *}_{12}(\mathscr{C}^{(i)}_+) \otimes \pi^{\circ *}_{23}(\mathscr{C}^{(j)}_+) \otimes \pi^{\circ *}_{31}(\mathscr{C}^{(k)}_+))$$

On the critical locus of $\pi^{\circ *}_{i,i+1}(W^\circ)$ we have $Y_i = \mathrm{Ad}_{g_i} Y_{i+1}$ hence on the critical locus the conjugation by g_1 intertwines the projections pr_1 and pr_2 the isomorphism (3.14) follows.

In the argument above we ignore the stability conditions but one can check that the shrinking Lemma 3.16 implies that the argument above works even after we impose the stability conditions.

3.7.2 *Second Markov Move*

The second Markov relation is more subtle and the proof of this relation is arguably the most valuable result of [26]. To convey the main idea of the proof we explain why it holds for the braids on two strands. In this case we need to compare the homology of the closure of $\sigma_1^{\pm 1}$ with the homology of unknot, so let us first do the most trivial case of the braids on one strands since $\mathrm{L}(\mathbf{1}_1)$, $\mathbf{1}_1 \in \mathfrak{Br}_1$ is manifestly the unknot.

Indeed, for $n = 1$ we have $\bar{\mathscr{X}_1} = \mathbb{C}\times\mathbb{C}^*\times 0$ and j_e embeds $\widetilde{\mathrm{FHilb}}_1^{free} = \mathbb{C}\times 1\times 0$ inside $\bar{\mathscr{X}_1}$. The group $B_1 = \mathbb{C}^*$ acts trivially on $\widetilde{\mathrm{FHilb}}_1^{free}$ and thus $\mathrm{FHilb}_1 = \mathbb{C}$ and $\mathbb{S}_e = j_e^*(\mathscr{O}_{\bar{\mathscr{X}_1}}) = \mathscr{O}_\mathbb{C}$ and $\mathscr{B}_1$ is the trivial bundle. We conclude then:

$$\dim_{q,t} H^0(\mathbf{1}_1) = \dim_{q,t} H^1(\mathbf{1}_1) = \frac{1}{1-q^2}.$$

Now let us explore the geometry of the free Hilbert scheme FHilb_2^{free}. Let us fix coordinates on the space $\widetilde{\mathrm{FHilb}}_2^{free} \subset \mathfrak{b}\times\mathfrak{n}\times V$:

$$X = \begin{bmatrix} x_{11} & x_{12} \\ 0 & x_{22} \end{bmatrix}, \quad Y = \begin{bmatrix} 0 & y \\ 0 & 0, \end{bmatrix}, \quad v = \begin{bmatrix} v_1 \\ v_2 \end{bmatrix}.$$

Since we have the stability condition $\mathbb{C}\langle X, Y\rangle v = \mathbb{C}^2$ and both X, Y are upper-triangular, we must have $v_2 \neq 0$. Thus after conjugating by the appropriate upper-triangular matrix we could assume that $v_2 = 1$, $v_1 = 0$, let us denote this vector by

v^0. It is also elementary to see that

$$\mathbb{C}\langle X, Y\rangle v^0 = \mathbb{C}^2 \text{ if and only if } x_{12}y \neq 0.$$

Also the stabilizer of v^0 is $\mathbb{C}^*$ that scales x_{12}, y and preserves x_{11}, x_{22}. Hence we have shown:

$$\mathrm{FHilb}_2^{free} = \mathbb{P}^1 \times \mathbb{C}^2,$$

the projection p on $\mathbb{C}^2$ is given by the coordinates x_{11}, x_{22}.

Let us contrast the geometry of FHilb_2^{free} with the geometry of FHilb_2. The discussion in this paragraph is not used in the proof below and is just an illustration of difficulties of the geometry of the flag Hilbert scheme. The condition $[X, Y] = 0$ is equivalent to the constraint:

$$y(x_{11} - x_{22}) = 0.$$

Hence the fibers of the projection $p : \mathrm{FHilb}_2 \to \mathbb{C}^2$ are points outside of the diagonal $x_{11} = x_{22}$ and the fibers are projective lines $\mathbb{P}^1$ over the diagonal.

Next let us recall that the matrix factorization for the simple positive crossing is $\mathscr{C}_+ = [\tilde{x}, yg_{21}]$. Since $\tilde{x}|_{g=1} = (x_{11} - x_{22})$, the pull-back $j_e(\mathscr{C}_+)$ is the Koszul complex that is homotopy equivalent to the structure sheaf of $\mathbb{P}^1 \times \mathbb{C}$. Finally, the tautological vector bundle is a sum of the line bundles $\mathscr{B}^\vee = \mathscr{O} \oplus \mathscr{O}(-1)$, hence:

$$\begin{aligned} H^0(\sigma_1) &= H^*(\mathscr{O}_{\mathbb{P}^1\times\mathbb{C}}) = \mathbb{C}[x_{11}], \\ H^1(\sigma_1) &= H^*(\mathscr{B}^\vee) = H^*(\mathscr{O}_{\mathbb{P}^1\times\mathbb{C}} \oplus \mathscr{O}_{\mathbb{P}^1\times\mathbb{C}}(-1)) = \mathbb{C}[x_{11}], \\ H^2(\sigma) &= H^*(\det(\mathscr{B})) = H^*(\mathscr{O}_{\mathbb{P}^1\times\mathbb{C}}(-1)) = 0. \end{aligned}$$

By our construction the matrix factorization for the negative crossing differs by a line bundle twist from the one for the positive crossing. In particular, we have $j_e^*(\mathscr{C}_-) = \mathscr{O}_{\mathbb{P}^1\times\mathbb{C}}(-1)$ and can compute the homology:

$$\begin{aligned} H^0(\sigma_1^{-1}) &= H^*(\mathscr{O}_{\mathbb{P}^1\times\mathbb{C}}(-1)) = 0, \\ H^2(\sigma^{-1}) &= H^*(\det(\mathscr{B}) \otimes \mathscr{O}(-1)) = H^*(\mathscr{O}_{\mathbb{P}^1\times\mathbb{C}}(-2)) = \mathbb{C}[x_{11}], \\ H^1(\sigma_1^{-1}) &= H^*(\mathscr{B}^\vee \otimes \mathscr{O}(-1)) = H^*(\mathscr{O}_{\mathbb{P}^1\times\mathbb{C}}(-1) \oplus \mathscr{O}_{\mathbb{P}^1\times\mathbb{C}}(-2)) = \mathbb{C}[x_{11}], \end{aligned}$$

Thus we have shown $H^k(\sigma) = H^k(\mathbf{1}_1)$ and $H^{k+1}(\sigma^{-1}) = H^k(\mathbf{1}_1)$ as we expected.

Respectively, we can use nested nature of the scheme FHilb_n to define the intermediate map:

$$\pi : \mathrm{FHilb}_n^{free} \to \mathbb{C} \times \mathrm{FHilb}_{n-1}^{free},$$

where the first component of the map π is x_{11} and the second component is just forgetting of the first rows and rows of the matrices X, Y and the first component of the vector v. Let us also fix notation for the line bundles on FHilb_n^{free}: we denote by $\mathscr{O}_k(-1)$ the line bundle induced from the twisted trivial bundle $\mathscr{O} \otimes \chi_k$. It is quite elementary to show

Proposition 3.25 *The fibers of the map π are projective spaces $\mathbb{P}^{n-1}$ and*

1. $\mathscr{B}_n/\pi^*(\mathscr{B}_{n-1}) = \mathscr{O}_n(-1)$.
2. $\mathscr{O}_n(-1)|_{\pi^{-1}(z)} = \mathscr{O}_{\mathbb{P}^{n-1}}(-1)$.

We can combine the last proposition with the observation that the total homology $H^*(\mathbb{P}^{n-1}, \mathscr{O}(-l))$ vanish if $l \in (1, n-1)$ and is one-dimensional for $l = 0, n$:

Corollary 3.26 *For any n we have:*

- $\pi_*(\Lambda^k \mathscr{B}_n) = \Lambda^k \mathscr{B}_{n-1}$
- $\pi_*(\mathscr{O}_n(m) \otimes \Lambda^k \mathscr{B}_n) = 0$ *if* $m \in [-n+2, -1]$.
- $\pi_*(\mathscr{O}_n(-n+1) \otimes \Lambda^k \mathscr{B}_n) = \Lambda^{k-1} \mathscr{B}_{n-1}[n]$

The geometric version of the Markov move is the following

Theorem 3.27 *For any $\beta \in \mathfrak{Br}_{n-1}$ we have*

$$\mathbb{H}^k(\beta \cdot \sigma_1) = \mathbb{H}^k(\beta), \quad \mathbb{H}^k(\beta \cdot \sigma_1) = \mathbb{H}^{k-1}(\beta).$$

Sketch of a proof The main technical component of the proof is the careful analysis of the matrix factorizations $\bar{\mathscr{C}}_{\beta \cdot \sigma^{\pm 1}} \mathrm{MF}(\bar{\mathscr{X}}_n, \bar{W})$. It is shown in [26] that this curved complex $\bar{\mathscr{C}}_{\beta \cdot \sigma_1^{\epsilon}}$ has form:

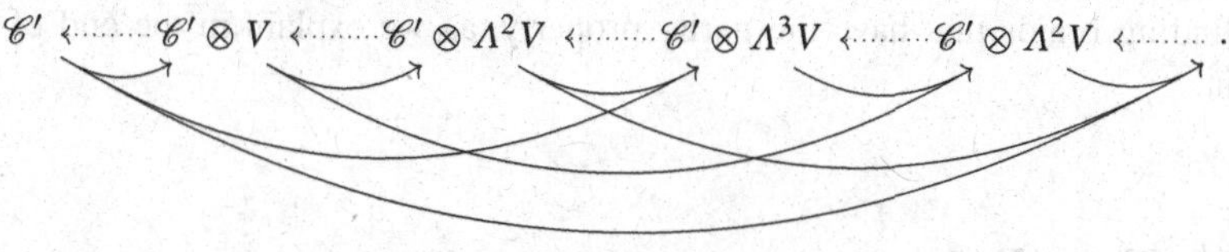

(3.15)

where $\mathscr{C}' = p_1^*(\bar{\mathscr{C}}_\beta)$, $V = \mathbb{C}^{n-2}$, the dotted arrows are the differentials of the Koszul complex for the ideal $I = (g_{13}, \ldots, g_{1n})$ where g_{ij} are the coordinates on the group inside the product $\bar{\mathscr{X}}_n = \mathfrak{b}_n \times \mathrm{GL}_n \times \mathfrak{n}_n$. Thus after the pull-back j_e^* the dotted arrows of the curved complex vanish and we only left with the arrows going from the left to right.

Now we would like to compute $\pi_*(j_e^*(\bar{\mathscr{C}}_{\beta \cdot \sigma_1^{\epsilon}}) \otimes \Lambda^k \mathscr{B}_n)$ and here we can apply the previous corollary. Thus if $\epsilon = 1$ then only the left extreme term of j_e^* of the complex (3.15) survive the push-forward π_*. Since the non-trivial arrows of j_e^* of (3.15) all are the solid arrows which are going the left to the right, the contraction of the π_*-acyclic terms do not lead to appearance of new correction arrows thus conclude that

$$\pi_*(j_e^*(\bar{\mathscr{C}}_{\beta\cdot\sigma_1}) \otimes \Lambda^k \mathscr{B}_n) = j_e^*(\bar{\mathscr{C}}_\beta \otimes \Lambda^k \mathscr{B}_{n-1}).$$

If $\epsilon = -1$ then only the right extreme term of j_e^* of the complex (3.15) survive the push-forward π_*. Hence the similar argument as before implies:

$$\pi_*(j_e^*(\bar{\mathscr{C}}_{\beta\cdot\sigma_1^{-1}}) \otimes \Lambda^k \mathscr{B}_n) = j_e^*(\bar{\mathscr{C}}_\beta \otimes \Lambda^{k-1} \mathscr{B}_{n-1}).$$

□

3.8 Chern Functor and Localization

Theorem 3.23 provides a theoretical method for constructing a sheaf on the flag Hilbert scheme that contains all the information about the knot homology of the closure of the braid L(β). However, it is hard to use this method for actually computing knot homology.

The first complication comes from the fact that the space FHilb_n is very singular and working with this space requires extra level of care and technicalities [11]. We will explain how one can circumvent this complication with the Chern functor from the next subsection.

The second complication comes from possible non-vanishing of the differential d in the theorem, one would like to avoid the spectral sequence that do not degenerate at the second step. The differential vanishes automatically if for example $\mathscr{S}_\beta^{odd}$ vanishes, this kind of property is probably related to the *parity* property in [7] for Soergel bimodel model of the knot homology. Again the Chern functor helps with finding braids that have the parity property, as we explain in the end of the section.

3.8.1 Chern Functor

In the paper [25] we construct a pair of functors which we call a Chern functor and a co-Chern functor:

$$\mathrm{MF}_n^{\mathrm{st}} \underset{\mathrm{HC}_{loc}^{st}}{\overset{\mathrm{CH}_{loc}^{st}}{\rightleftarrows}} D_{T_{sc}}^{per}(\mathrm{Hilb}_n)\ , \tag{3.16}$$

where Hilb_n is the Hilbert scheme of n points on $\mathbb{C}^2$, while $D^{per}_{T_{sc}}(\mathrm{Hilb}_n)$ is the derived category of two-periodic T_{sc}-equivariant complexes on the Hilbert scheme. In the same paper we prove

Theorem 3.28 ([25]) *For every n we have*

- *The functors* $\mathrm{CH}^{\mathrm{st}}_{\mathrm{loc}}$ *and* $\mathrm{HC}^{\mathrm{st}}_{\mathrm{loc}}$ *are adjoint.*
- *The functor* $\mathrm{HC}^{\mathrm{st}}_{\mathrm{loc}}$ *is monoidal.*
- *The image of* $\mathrm{HC}^{\mathrm{st}}_{\mathrm{loc}}$ *commutes with the elements* $\Phi(\beta)$, $\beta \in \mathfrak{Br}_n$.

As a manifestation of the categorified Riemann-Roch formula, we obtain a new interpretation for the triply-graded homology:

Theorem 3.29 ([25]) *For any* $\beta \in \mathfrak{Br}_n$ *we have:*

$$\mathrm{HHH}(\beta) = \mathrm{Hom}(\mathscr{O}, \mathrm{CH}^{\mathrm{st}}_{\mathrm{loc}}(\Phi(\beta)) \otimes \Lambda^\bullet \mathscr{B}).$$

Let us outline the construction of the Chern functor in the next subsection.

3.8.2 *Construction of* CH

First we will construct the functor between the categories MF and $\mathrm{MF}_{\mathrm{Dr}}$ where the last category is defined as a stable version of the category of equivariant matrix factorizations:

$$\mathrm{MF}_{\mathrm{Dr}} := \mathrm{MF}_G(\mathscr{C}, W_{\mathrm{Dr}}), \quad \mathscr{C} = \mathfrak{g} \times G \times \mathfrak{g}, \quad W_{\mathrm{Dr}}(Z, g, X) = \mathrm{Tr}(X(Z - \mathrm{Ad}_g Z)),$$

the group GL_n acting on components of $\mathscr{C}$ by conjugation. The stable version of the category is defined as category of matrix factorizations on the slightly enlarged space:

$$\mathscr{C}^{st} \subset \mathscr{C} \times V, \quad (Z, g, X, v) \in \mathscr{C}^{st} \text{ iff } gv = v \text{ and } \mathbb{C}\langle X, \mathrm{Ad}_g Z\rangle v = \mathbb{C}^n.$$

Both stable and unstable versions of the categories fit into the diagram:

$$\mathrm{MF}^\bullet \underset{\mathrm{HC}^\bullet}{\overset{\mathrm{CH}^\bullet}{\rightleftarrows}} \mathrm{MF}^\bullet_{\mathrm{Dr}},$$

where $\bullet$ can be either st or $\emptyset$.

To lighten the exposition we explain only the construction for the functors CH and HC, the stable version is an easy modification of the construction, see [25].

We need two auxiliary spaces in order to define the Chern functor:

$$\mathscr{Z}^0_{\mathrm{CH}} = \mathfrak{g} \times G \times \mathfrak{g} \times G \times \mathfrak{n}, \quad \mathscr{Z}_{\mathrm{CH}} = \mathfrak{g} \times G \times \mathfrak{g} \times G \times \mathfrak{b}$$

The action of $G \times B$ on these spaces is

$$(k, b) \cdot (Z, g, X, h, Y) = (\mathrm{Ad}_k(Z), \mathrm{Ad}_k(g), \mathrm{Ad}_k(X), khb, \mathrm{Ad}_{b^{-1}}(Y))$$

and the invariant potential is

$$W_{\mathrm{CH}}(Z, g, X, h, Y) = \mathrm{Tr}(X(\mathrm{Ad}_{gh}(Y) - \mathrm{Ad}_h(Y))).$$

The spaces $\mathscr{C}$ and $\mathscr{X}$ are endowed with the standard $G \times B^2$-equivariant structure, the action of B^2 on $\mathscr{C}$ is trivial. The following maps

$$\pi_{\mathrm{Dr}} : \mathscr{Z}_{\mathrm{CH}} \to \mathscr{C}, \quad f_\Delta : \mathscr{Z}^0_{\mathrm{CH}} \to \mathscr{X} \quad j^0 : \mathscr{Z}^0_{\mathrm{CH}} \to \mathscr{Z}_{\mathrm{CH}}.$$

$$\pi_{\mathrm{Dr}}(Z, g, X, h, Y) = (Z, g, X), \quad f_\Delta(Z, g, X, h, Y) = (X, gh, Y, h, Y)$$

are fully equivariant if we restrict the B^2-equivariant structure on $\mathscr{X}$ to the B-equivariant structure via the diagonal embedding $\Delta : B \to B^2$. Note that j^0 is an inclusion map.

The kernel of the Fourier-Mukai transform is the Koszul matrix factorization

$$\mathrm{K}_{\mathrm{CH}} := [X - \mathrm{Ad}_{g^{-1}}X, \mathrm{Ad}_h Y - Z] \in \mathrm{MF}(\mathscr{Z}_{\mathrm{CH}}, \pi^*_{\mathrm{Dr}}(W_{\mathrm{Dr}}) - f^*_\Delta(W)).$$

and we define the Chern functor:

$$\mathrm{CH}(\mathscr{C}) := \pi_{\mathrm{Dr}*}(\mathrm{CE}_{\mathfrak{n}}(\mathrm{K}_{\mathrm{CH}} \otimes (j^0_* \circ f^*_\Delta(\mathscr{C})))^T). \tag{3.17}$$

We also define the co-Chern functor HC as the adjoint functor that goes in the opposite direction: $\mathrm{HC}\colon \mathrm{MF}_{\mathrm{Dr}} \to \mathrm{MF}$. Thus, the functor HC is the composition of adjoints of all the functors that appear in the formula (3.17).

The product $\mathscr{Z}_{\mathrm{CH}} \times B$ has a $B \times B$-equivariant structure: for $(p, g) \in \mathscr{Z}_{\mathrm{CH}} \times B$ we define

$$(h_1, h_2) \cdot (p, g) = (h_1 \cdot p, h_1 g h_2^{-1})$$

Then the following map is B^2-equivariant:

$$\tilde{f}_\Delta : \mathscr{Z}^0_{\mathrm{CH}} \times B \to \mathscr{X} \times B,$$

$$\tilde{f}_\Delta(Z, g, X, h, Y, b) = (X, gh, Y, hb, \mathrm{Ad}_b Y, b).$$

The map $\tilde{f}_\Delta$ is a composition of the projection along the first factor of $\mathscr{Z}_{\mathrm{CH}}$ and the embedding inside $\mathscr{X} \times B$. The embedding is defined by the formula

$$\mathrm{Ad}_b Y_1 = Y_2,$$

so it is a regular embedding. Thus since

$$j^{0*}(\mathrm{K}_{\mathrm{CH}} \otimes \tilde{\pi}^*_{\mathrm{Dr}}(\mathscr{D})) \in \mathrm{MF}_{G\times B^2}(\mathscr{Z}_{\mathrm{CH}} \times B, \tilde{f}^*_\Delta(W)),$$

where $\tilde{\pi}_{\mathrm{Dr}} : \mathscr{Z}_{\mathrm{HC}} \times B \to \mathscr{C}$ is a natural extension of map π_{Dr} by the projection along B, we have a well-defined matrix factorization $\tilde{f}_{\Delta *} \circ j^{0*}(\mathrm{K}_{\mathrm{CH}} \otimes \pi^*_{\mathrm{Dr}}(\mathscr{D})) \in \mathrm{MF}_{G\times B^2}(\mathscr{X} \times B, \pi^*_B(W))$, where π_B is the projection along the last factor. Now we can define:

$$\mathrm{HC}(\mathscr{D}) := \pi_{B*}(\tilde{f}_{\Delta *} \circ j^{0*}(\mathrm{K}_{\mathrm{CH}} \otimes \pi^*_{\mathrm{Dr}}(\mathscr{D}))). \tag{3.18}$$

3.8.3 Linear Koszul Duality

We need to relate the category $\mathrm{MF}^{st}_{\mathrm{Dr}}$ and the category $D^{per}_{T_{sc}}(\mathrm{Hilb})$. This relation is a particular example of the linear Koszul duality. Let us discuss the linear Koszul duality in general.

Derived algebraic geometry is explained in many places, here we explain it in the most elementary setting sufficient for our needs.

Initial data for an affine derived complete intersection is a collection of elements $f_1, \dots, f_m \in \mathbb{C}[X]$. It determines the differential graded algebra

$$\mathscr{R} = (\mathbb{C}[X] \otimes \Lambda^* U, D), \quad D = \sum_{i=1}^m f_i \frac{\partial}{\partial \theta_i},$$

where θ_i from a basis of $U = \mathbb{C}^m$.

More generally, given a dg algebra $\mathscr{R}$ such that $H^0(\mathscr{R}) = \mathscr{O}_Z$ we say that $Spec(\mathscr{R})$ is a dg scheme with underlying scheme Z. Respectively, we define dg category of coherent sheaves on $Spec(\mathscr{R})$ as

$$\mathrm{Coh}(Spec(\mathscr{R})) = \frac{\{\text{bounded complexes of finitely generated } \mathscr{R} \text{ dg modules}\}}{\{\text{quasi-isomorphisms}\}}.$$

Consider a potential on $X \times U$:

$$W = \sum_{i=1}^m f_i(x) z_i,$$

where z_i is a basis of U^* dual to the basis θ_i. For the Koszul matrix factorization:

$$\mathrm{MF}(X \times U, W) \ni \mathrm{B} = (\mathscr{R} \otimes \mathbb{C}[U], D_B), \quad D_B = \sum_{i=1}^{m} z_i \theta_i + f_i \frac{\partial}{\partial \theta_i}.$$

and for a (M, D_M) dg module over $\mathscr{R}$, the tensor product

$$\mathrm{KSZ}_U(M) := M \otimes_{\mathbb{C}[X] \otimes \Lambda^*(U)} \mathrm{B}$$

is an object of $\mathrm{MF}(X \times U, W)$ with the differential $D = D_M \otimes 1 + 1 \otimes D_B$. The map KSZ_U extends to a functor between triangulated categories:

$$\mathrm{KSZ}_U : \mathrm{Coh}(Spec\ (\mathscr{R})) \to \mathrm{MF}(X \times U, W).$$

The functor in the other direction is based on the dual matrix factorization:

$$\mathrm{MF}(X \times U, -W) \ni B^* = (\mathscr{R} \otimes \mathbb{C}[U], D_B^*), \quad D_B = \sum_{i=1}^{m} z_i \theta_i - f_i \frac{\partial}{\partial \theta_i},$$

$$\mathrm{KSZ}_U^* : \mathrm{MF}(X \times U, W) \to \mathrm{Coh}(Spec\ (\mathscr{R})),$$
$$\mathrm{KSZ}_U^*(\mathscr{F}) := \mathrm{Hom}_{\mathscr{R}}(\mathscr{F} \otimes_{\mathbb{C}[X \times U]} \mathrm{B}^*, \mathscr{R}).$$

Theorem 3.30 *The compositions of the functors:*

$$\mathrm{KSZ}_U \circ \mathrm{KSZ}_U^*, \quad \mathrm{KSZ}_U^* \circ \mathrm{KSZ}_U$$

are autoequivalences of the corresponding categories.

Proof of this theorem could be found in [1, 14] or one can consult [25] for a more streamlined argument.

3.8.4 Linearization

We would like to apply the linear Koszul in our situation. The complication in our case is that we want to eliminate the group factor in the space $\mathscr{C}^{st}$ but the group is not a linear space. Thus we have to restrict ourselves to the neighborhood of the identity and linearize the potential in this neighborhood and as we explain below it could be done with localization.

A coordinate substitution $Y = Ug^{-1}$ on our main variety $\mathscr{C}$ makes the potential tri-linear:

$$\underline{W}_{\mathrm{Dr}}(X, U, g) = \mathrm{Tr}(X[U, g]) = W_{\mathrm{Dr}}(X, Ug^{-1}, g).$$

Thus we introduce linearized categories:

$$\underline{\mathrm{MF}}^{\bullet}_{\mathrm{Dr}} := \mathrm{MF}_G(\underline{\mathscr{C}}^{\bullet}, \underline{W}_{\mathrm{lin}}),$$

where $\underline{\mathscr{C}}^{\bullet}$ is obtained from $\mathscr{C}^{\bullet}$ by taking the closure of G inside $\mathfrak{g}$.

Since $j_G \colon \mathscr{C}^{\bullet} \hookrightarrow \underline{\mathscr{C}}^{\bullet}$ is an open embedding, the pull-back functor j_G^* is a localization functor and we denote

$$\mathrm{loc}^{\bullet} \colon \mathrm{MF}^{\bullet}_{\mathrm{Dr}} \to \underline{\mathrm{MF}}^{\bullet}_{\mathrm{Dr}}$$

for this functor.

Proposition 3.31 ([25]) *The functors* $\mathrm{loc}^{\mathrm{st}}$ *are isomorphisms.*

Since the potential $\underline{W}$ is linear as a function of $g \in \mathfrak{g}$ and the scaling torus T_{sc} does not act on g, we obtain a pair of mutually inverse functors:

$$\underline{\mathrm{MF}}^{\bullet}_{\mathrm{Dr}} \underset{\mathrm{KSZ}_{\mathfrak{g}}}{\overset{\mathrm{KSZ}^*_{\mathfrak{g}}}{\rightleftarrows}} \mathrm{Coh}^{\bullet}$$

here Coh^{st} is the two-periodic derived category $D^{per}(\mathrm{Hilb})$ and Coh is the DG category of the commuting variety.

The functors that we wanted to construct are defined by the composing the functors:

$$\mathrm{CH}^{\mathrm{st}}_{\mathrm{loc}} := \mathrm{CH}^{\mathrm{st}} \circ (\mathrm{loc}^{\mathrm{st}})^{-1} \circ \mathrm{KSZ}^*_{\mathfrak{g}} : \mathrm{MF}^{\bullet} \to \mathrm{D}^{\mathrm{per}}(\mathrm{Hilb}_n(\mathbb{C}^2)).$$

The localization functor does not seem to be invertible in the case of $\bullet = \emptyset$, however a construction of the functor in the opposite direction does not require invertibility of the localization:

$$\mathrm{HC}^{\bullet}_{\mathrm{loc}} := \mathrm{HC}^{\bullet} \circ \mathrm{loc}^{\bullet} \circ \mathrm{KSZ}_{\mathfrak{g}} \colon \mathrm{Coh}^{\bullet} \to \mathrm{MF}^{\bullet}.$$

3.8.5 Localization Formulas

The advantage of this new interpretation is that the Hilbert scheme is smooth, unlike the flag Hilbert scheme which is a homological support of $\mathscr{E}\mathrm{xt}(\Phi(\beta), \Phi(1))$. So the complexes on Hilb are more manageable than their flag counter-part. In support of this expectation, we apply the Chern functor to the Jusys-Murphy (JM) subgroup inside $\mathfrak{Br}_n$ together with the parity property and prove an explicit localization formula for the sufficiently positive elements of the JM subgroup.

Recall that the JM subgroup is generated by the elements

$$\delta_i = \sigma_i\sigma_{i+1}\dots\sigma_{n-1}^2\dots\sigma_{i+1}\sigma_i.$$

It is not hard to see that these elements mutually commute and that the full twist from the introduction is the product:

$$\mathrm{FT} = \prod_{i=1}^{n-1}\delta_i.$$

It is expected that $\mathrm{CH}^{\mathrm{st}}_{\mathrm{loc}}$ applied to the matrix factorization corresponding to the sufficiently positive element of JM algebra is a sheaf supported in one homological degree, we state the precise conjecture below. Modulo this geometric conjecture we have a (conditional on the conjecture) formula for the corresponding homology of the links.

Theorem* 3.32 *For any n there are $N, M > 0$ such that for a vector $\mathbf{b} \in \mathbb{Z}^{n-1}$ with $a_{i+1} - a_i > N, a_2 > M$ the (q,t,a)-character of the homology of the closure of the braid $\prod_{i=2}\delta^{b_i}$ is given by the formula*

$$\dim_{a,Q,T}\mathrm{HHH}(\prod_{i=2}^{n}\delta^{b_i}) = \sum_T\prod_i\frac{z_i^{b_i}(1+az_i^{-1})}{1-z^{-1}}\prod_{1\le i<j\le n}\zeta(\frac{z_i}{z_j}),$$

where $\zeta(x) = \frac{(1-x)(1-QTx)}{(1-Qx)(1-Tx)}$, $Q = q^2, T = t^2/q^2$. The last sum is over all standard Young tableaux with $z_i = Q^{a'(i)}T^{l'(i)}$, a', l' are co-arm and co-leg of the square the standard tableau with the square with the label i.

The proof has two components. The first component is concerned with actual computation of the matrix factorization $\Phi(\prod_{i=2}\delta_i^{b_i})$. This computation is an easy consequence of our construction of Φ^{aff}:

Theorem 3.33 ([24]) *For any $i = 1, \dots, n$ we have*

$$\Phi^{aff}(\Delta_i) = \Phi^{aff}(1)\langle\chi_i, 0\rangle.$$

In particular, we show in [24] that the pull-back j_{st}^* sends $\Phi^{aff}(1)\langle\chi_n, 0\rangle$ to the trivial line bundle. Since $\delta_i = \mathrm{fgt}(\Delta_i)$ we conclude the following

Corollary 3.34 *For any $\beta \in \mathfrak{Br}_n$ and $b_i, M \in \mathbb{Z}$ we have*

$$\Phi(\beta \cdot \prod_{i=2}^{n} \delta_i^{b_i}) = \Phi(\beta)\langle \mathbf{b}, 0\rangle,$$

$$\mathrm{CH}^{\mathrm{st}}_{\mathrm{loc}}(\Phi(\beta \cdot \mathrm{FT}^M)) = \mathrm{CH}^{\mathrm{st}}_{\mathrm{loc}}(\Phi(\beta)) \otimes \det(\mathscr{B})^{\otimes M},$$

where $\mathrm{FT} = \prod_{i=2}^{n} \delta_i$ *is the full-twist braid.*

Thus we can apply the first formula from the corollary and get an explicit Koszul matrix factorization describing the desired curved complex:

$$\Phi(\prod_{i=2}^{n} \delta_i^{b_i}) = \mathscr{C}_{\|}\langle \mathbf{b}, 0\rangle.$$

It is much harder though to compute Chern functor $\mathrm{CH}(\mathscr{C}_{\|}\langle \mathbf{b}, 0\rangle)$. It is expected [11] that $\mathrm{CH}(\mathscr{C}_{\|})$ is a celebrated Procesi vector bundle and finding an explicit description for this vector bundle is notoriously hard [12]. So at the moment we do not have an explicit statement for $\mathrm{CH}(\mathscr{C}_{\|}\langle \mathbf{b}, 0\rangle)$ but we believe the following weaker conjecture could be proved by inductive argument from the work of Haiman.

Conjecture 3.35 ([27]) There is $N > 0$ such that for any $\mathbf{a}$ such that $a_{i+1} - a_i > N$ the two periodic complex $\mathrm{CH}^{\mathrm{st}}_{\mathrm{loc}}(\mathscr{C}_{\|}\langle \mathbf{a}, 0\rangle)$ is homotopy equivalent to the sheaf concentrated in even homological degree.

On the other hand $\det(\mathscr{B})$ is an ample line bundle on $\mathrm{Hilb}_n(\mathbb{C}^2)$ and hence the assumptions of the Theorem 3.32 and Corollary 3.34 imply that $\mathrm{CH}^{\mathrm{st}}_{\mathrm{loc}}(\mathscr{C}_{\|}\langle \mathbf{b}\rangle)$ is homotopy equivalent to the sheaf with no higher homology. The differential in the complex $\mathscr{C}_{\|}$ has T_{sc} degree t, respectively by t-twisting even component of $\mathscr{C}_{\|}$ we obtain the curved complex $\mathscr{C}_{\|}^{ev}$ with T_{sc}-invariant differential. From the discussion above we have:

$$H^*(\mathrm{CH}^{\mathrm{st}}_{\mathrm{loc}}(\mathscr{C}_{\|}\langle \mathbf{a}, 0\rangle)) = H^0(\mathrm{CH}^{\mathrm{st}}_{\mathrm{loc}}(\mathscr{C}_{\|}^{ev}\langle \mathbf{a}, 0\rangle)) = \chi(\mathrm{CH}^{\mathrm{st}}_{\mathrm{loc}}(\mathscr{C}_{\|}^{ev}\langle \mathbf{a}, 0\rangle))$$
$$= \chi(\mathbb{S}_1^{ev}\langle \mathbf{a}, 0\rangle),$$

where $\mathbb{S}_1^{ev}$ is the version of $\mathbb{S}_1$ with t-twisted even component. There is well-defined image $\mathrm{K}(\mathscr{C}_{\|})$ of the complex inside of the T_{sc}-equivariant K-theory. Thus the Euler characteristics of the LHS of last formula can be computed within $K_{T_{sc}}(\mathrm{FHilb}^{free})$ and here we can use the analog Negut's theorem for the push-forward along the fibers of the projection

Proposition 3.36 *For any rational function $r(\mathscr{L}_n)$ of $\mathscr{L}_n = \mathscr{O}_n(-1)$ with coefficients rational functions of $\mathscr{L}_i = \mathscr{O}_i(-1)$, $i < n$, the K-theory push-forward is given by*

$$\pi_*(r(\mathscr{L}_n)) = \int \frac{r(z)}{(1 - z^{-1})} \prod_{i=1} \zeta'(\mathscr{L}_i/z) \frac{dz}{z}$$

where the contour of integration separates the set $\mathrm{Poles}(r(z)) \bigcup \{0, \infty\}$ *from the poles of the rest of the integrant.*

The K-theory class of the complex $\mathrm{CH}^{\mathrm{st}}_{\mathrm{loc}}(\mathscr{C}^{ev}_{\|})$ is $\prod_{1 \le i < j \le n}(1 - qt\mathscr{L}_i/\mathscr{L}_j)$. Hence we can apply the formula from the previous proposition iterative to obtain the iterated residue integral formula for the desired link invariant:

$$\int \cdots \int \prod_i \frac{z_i^{b_i}(1 + az_i^{-1})}{1 - z^{-1}} \prod_{1 \le i < j \le n} \zeta(\frac{z_i}{z_j}) \frac{dz_1}{z_1} \cdots \frac{dz_n}{z_n}.$$

The final step of the proof is a delicate analysis of the iterated residue that was done in the work of Negut [20] in the context of K-theory of the flag Hilbert scheme.

Acknowledgements First of all I would like to thank my coathor and friend Lev Rozansky for teaching everything that is in these notes. All results in these notes are contained in our joint papers. I also would like to thank Andrei Neguţ and Tina Kanstrup for discussion related to the content of the notes. I am very grateful to an anonymous referee for many great suggestion on improving the first draft of the notes. I am very grateful to CIME foundation for opportunity to teach the course at the Summer school. The participants of the course provided valuable feed-back on the material. I was also partially supported in part by the NSF CAREER grant DMS-1352398, the NSF FRG grant DMS-1760373 grant and Simons Foundation Fellowship No.561855.

References

1. S. Arkhipov, T. Kanstrup, Braid group actions on matrix factorizations (2015). http://arxiv.org/abs/1510.07588
2. S. Arkhipov, T. Kanstrup, Equivariant matrix factorizations and hamiltonian reduction. Bull. Korean Math. Soc. (5), 1803–1825 (2015). http://arxiv.org/abs/1510.07472v1
3. N. Chris, V. Ginzburg, *Representation Theory and Complex Geometry* (Birkhauser, Basel, 1997)
4. D. Eisenbud, Homological algebra on a complete intersection, with an application to group representations. Trans. Am. Math. Soc. **260**(1), 35–64 (1980)
5. T. Dyckerhoff, Compact generators in categories of matrix factorizations. Duke Math. J. **159**(2):223–274, 2011.
6. B. Elias, M. Hogancamp, On the computation of torus link homology (2016). http://arxiv.org/abs/1603.00407
7. B. Elias, M. Hogancamp, Categorical diagonalization (2017). http://arxiv.org/abs/1707.04349v1
8. B. Elias, M. Hogancamp. Categorical diagonalization of full twists (2017). http://arxiv.org/abs/1801.00191v1
9. E. Gorsky, M. Hogancamp, Hilbert schemes and y-ification of Khovanov-Rozansky homology (2017). http://arxiv.org/abs/1712.03938v1
10. E. Gorsky, A. Oblomkov, J. Rasmussen, V. Shende, Torus knots and the rational DAHA. Duke Math. J. **163**, 2709–2794 (2014)
11. E. Gorsky, J. Rasmussen, A. Negut, Flag Hilbert schemes, colored projectors and Khovanov-Rozansky homology (2016). http://arxiv.org/abs/1608.07308
12. M. Haiman. Vanishing theorems and character formulas for the Hilbert scheme of points in the plane. Invent. Math. **149**(2), 371–407 (2002)

13. D. Halpern-Leistner (2019, in preparation)
14. U. Isik, Equivalence of the derived category of a variety with a singularity category. Int. Math. Res. Not. **2013**(12), 2728–2808 (2013). http://arxiv.org/abs/1011.1484v1
15. V.F.R. Jones, Hecke algebra representations of braid groups and link polynomials. Ann. Math. **126**(2), 335 (1987). https://doi.org/10.2307/1971403
16. A. Kapustin, L. Rozansky, Three-dimensional topological field theory and symplectic algebraic geometry II. Commun. Number Theory Phys. **4**(3), 463–549 (2010). https://doi.org/10.4310/cntp.2010.v4.n3.a1.
17. A. Kapustin, L. Rozansky, N. Saulina, Three-dimensional topological field theory and symplectic algebraic geometry I. Nucl. Phys. B **816**(3), 295–355 (2009). https://doi.org/10.1016/j.nuclphysb.2009.01.027
18. M. Khovanov, A categorification of the Jones polynomial. Duke Math. J. **101**(3), 359–426 (2000) https://doi.org/10.1215/s0012-7094-00-10131-7.
19. M. Khovanov, L. Rozansky, Matrix factorizations and link homology II. Geom. Topol. **12**, 1387–1425 (2008)
20. A. Negut. Moduli of flags of sheaves and their K-theory. Algebr. Geom. **2**(1), 19–43 (2015) https://doi.org/10.14231/ag-2015-002.
21. A. Oblomkov, L. Rozansky, HOMFLYPT homology of Coxeter links (2017). http://arxiv.org/abs/1706.00124v1
22. A. Oblomkov, L. Rozansky, 3D TQFT and HOMFLYPT homology (2018). http://arxiv.org/abs/1812.06340v1
23. A. Oblomkov, L. Rozansky, A categorification of a cyclotomic Hecke algebra (2018). http://arxiv.org/abs/1801.06201v1
24. A. Oblomkov, L. Rozansky, Affine braid group, JM elements and knot homology. Transform. Groups (2018). https://doi.org/10.1007/s00031-018-9478-5
25. A. Oblomkov, L. Rozansky, Categorical Chern character and braid groups (2018). http://arxiv.org/abs/1811.03257v1
26. A. Oblomkov, L. Rozansky, Knot homology and sheaves on the Hilbert scheme of points on the plane. Sel. Math. **24**(3), 2351–2454 (2018). https://doi.org/10.1007/s00029-017-0385-8
27. A. Oblomkov, L. Rozasky, Categorical Chern character and Hall algebras (2018, in preparation)
28. A. Oblomkov, J. Rasmussen, V. Shende, The Hilbert scheme of a plane curve singularity and the HOMFLY homology of its link. Geom. Topol. **22**(2), 645–691 (2018). https://doi.org/10.2140/gt.2018.22.645
29. D. Orlov, Triangulated categories of singularities and D-branes in Landau-Ginzburg models. Proc. Steklov Inst. Math. **246**(3), 227–248 (2004)
30. D. Orlov, Derived categories of coherent sheaves and triangulated categories of singularities. Algebra Arithmetic Geometry, 503–531 (2009). https://doi.org/10.1007/978-0-8176-4747-6_16
31. D. Orlov, Matrix factorizations for nonaffine LG-models. Math. Ann. **353**(1), 95–108 (2011) https://doi.org/10.1007/s00208-011-0676-x
32. A. Polishchuk, A. Vaintrob, Matrix factorizations and singularity categories for stacks. Ann. Inst. Fourier **61**(7), 2609–2642 (2011). https://doi.org/10.5802/aif.2788
33. W. Soergel, Langlands' philosophy and Koszul duality, in *Algebra-Representation Theory (Constanta,2000)*, pp. 379–414 (2001)

LECTURE NOTES IN MATHEMATICS

Springer

Editorial Policy

1. Lecture Notes aim to report new developments in all areas of mathematics and their applications – quickly, informally and at a high level. Mathematical texts analysing new developments in modelling and numerical simulation are welcome.

   Manuscripts should be reasonably self-contained and rounded off. Thus they may, and often will, present not only results of the author but also related work by other people. They may be based on specialised lecture courses. Furthermore, the manuscripts should provide sufficient motivation, examples and applications. This clearly distinguishes Lecture Notes from journal articles or technical reports which normally are very concise. Articles intended for a journal but too long to be accepted by most journals, usually do not have this "lecture notes" character. For similar reasons it is unusual for doctoral theses to be accepted for the Lecture Notes series, though habilitation theses may be appropriate.

2. Besides monographs, multi-author manuscripts resulting from SUMMER SCHOOLS or similar INTENSIVE COURSES are welcome, provided their objective was held to present an active mathematical topic to an audience at the beginning or intermediate graduate level (a list of participants should be provided).

   The resulting manuscript should not be just a collection of course notes, but should require advance planning and coordination among the main lecturers. The subject matter should dictate the structure of the book. This structure should be motivated and explained in a scientific introduction, and the notation, references, index and formulation of results should be, if possible, unified by the editors. Each contribution should have an abstract and an introduction referring to the other contributions. In other words, more preparatory work must go into a multi-authored volume than simply assembling a disparate collection of papers, communicated at the event.

3. Manuscripts should be submitted either online at www.editorialmanager.com/lnm to Springer's mathematics editorial in Heidelberg, or electronically to one of the series editors. Authors should be aware that incomplete or insufficiently close-to-final manuscripts almost always result in longer refereeing times and nevertheless unclear referees' recommendations, making further refereeing of a final draft necessary. The strict minimum amount of material that will be considered should include a detailed outline describing the planned contents of each chapter, a bibliography and several sample chapters. Parallel submission of a manuscript to another publisher while under consideration for LNM is not acceptable and can lead to rejection.

4. In general, **monographs** will be sent out to at least 2 external referees for evaluation.

   A final decision to publish can be made only on the basis of the complete manuscript, however a refereeing process leading to a preliminary decision can be based on a pre-final or incomplete manuscript.

   Volume Editors of **multi-author works** are expected to arrange for the refereeing, to the usual scientific standards, of the individual contributions. If the resulting reports can be forwarded to the LNM Editorial Board, this is very helpful. If no reports are forwarded or if other questions remain unclear in respect of homogeneity etc, the series editors may wish to consult external referees for an overall evaluation of the volume.

5. Manuscripts should in general be submitted in English. Final manuscripts should contain at least 100 pages of mathematical text and should always include
   - a table of contents;
   - an informative introduction, with adequate motivation and perhaps some historical remarks: it should be accessible to a reader not intimately familiar with the topic treated;
   - a subject index: as a rule this is genuinely helpful for the reader.
   - For evaluation purposes, manuscripts should be submitted as pdf files.

6. Careful preparation of the manuscripts will help keep production time short besides ensuring satisfactory appearance of the finished book in print and online. After acceptance of the manuscript authors will be asked to prepare the final LaTeX source files (see LaTeX templates online: https://www.springer.com/gb/authors-editors/book-authors-editors/manuscriptpreparation/5636) plus the corresponding pdf- or zipped ps-file. The LaTeX source files are essential for producing the full-text online version of the book, see http://link.springer.com/bookseries/304 for the existing online volumes of LNM). The technical production of a Lecture Notes volume takes approximately 12 weeks. Additional instructions, if necessary, are available on request from lnm@springer.com.

7. Authors receive a total of 30 free copies of their volume and free access to their book on SpringerLink, but no royalties. They are entitled to a discount of 33.3 % on the price of Springer books purchased for their personal use, if ordering directly from Springer.

8. Commitment to publish is made by a *Publishing Agreement*; contributing authors of multiauthor books are requested to sign a *Consent to Publish form*. Springer-Verlag registers the copyright for each volume. Authors are free to reuse material contained in their LNM volumes in later publications: a brief written (or e-mail) request for formal permission is sufficient.

Addresses:
Professor Jean-Michel Morel, CMLA, École Normale Supérieure de Cachan, France
E-mail: moreljeanmichel@gmail.com

Professor Bernard Teissier, Equipe Géométrie et Dynamique,
Institut de Mathématiques de Jussieu – Paris Rive Gauche, Paris, France
E-mail: bernard.teissier@imj-prg.fr

Springer: Ute McCrory, Mathematics, Heidelberg, Germany,
E-mail: lnm@springer.com

Printed in the United States
By Bookmasters